AI-Driven Cybersecurity Tech Marketing: Future-Proof Strategies for Sales, Success and AI-Powered Growth

Ready-to-Use Creative Ideas, Clever Tactics, and AI-Driven Implementation for Marketers in High Tech and Cybersecurity

---

By Nikolay Gul
AI-Driven Marketing Strategist & Cybersecurity Branding Expert

Connect with the author, Nikolay Gul,
on LinkedIn for more insights and updates:
https://www.linkedin.com/in/webdesignerny/

LCCN (Library of Congress Control Number):
2025902819
ISBN: 9798218612481

Welcome to the Future of Cybersecurity and High-Tech Marketing

Marketing in cybersecurity and high-tech industries has always required a balance of trust, precision, and technological insight. While traditional strategies have served well in the past, new AI-driven tools and data-backed approaches offer exciting opportunities to enhance, optimize, and refine how companies connect with their audience.

This book is designed to help you adapt, innovate, and stay ahead in an industry where trust is essential, and cutting-edge strategies make the difference.

I'm Nikolay Gul, and through my experience working with MSPs, SaaS vendors, and cybersecurity companies, I've seen firsthand how effective marketing can drive business growth. This book isn't about discarding what works it's about enhancing your marketing with fresh perspectives, AI-powered tools, and creative strategies that can make a meaningful impact.

This is not another theoretical marketing book. It's a field-tested guide, packed with:

- 100+ AI-powered marketing prompts to spark new ideas
- Data-driven strategies inspired by real cybersecurity events (like the U.S. Treasury hack).
- Sales and lead-generation techniques that align with high-tech decision-makers' expectations.
- Step-by-step frameworks for leveraging AI in branding, content marketing, and automation.

Why This Book?

I've spent years helping MSPs, SaaS vendors, and cybersecurity companies stand out in crowded markets.

This book condenses that experience into actionable, ready-to-implement strategies that will:

- Boost your authority and industry credibility
- Attract high-value clients through trust-based marketing
- Leverage AI for smarter, faster marketing execution

By the time you finish this book, you'll gain valuable insights and tools to enhance your approach to cybersecurity marketing, regardless of your technical background.

What You'll Gain from This Book

This book is structured as a **practical playbook**, giving you tools to:

- Create custom AI-generated marketing content in minutes
- Develop **industry-specific campaigns** that align with real-world cybersecurity trends
- Automate repetitive marketing tasks so you can focus on strategy and creativity
- Understand how to apply psychological triggers (hooks) in cybersecurity sales and marketing

The Power of Collaboration

Marketing doesn't work in isolation. The most successful cybersecurity and high-tech companies thrive when **marketing and sales teams work in sync**, aligning messaging, sharing insights, and using AI-driven tools to enhance collaboration. This book provides strategies to bridge that gap—so that your company's growth isn't just about clever ideas, but about executing them effectively.

The Creative Edge in Cybersecurity and High-Tech Marketing

Some marketing books treat cybersecurity and high-tech like other industry. But they're not. You're not just selling a product or service; you're selling peace of mind, trust, and the promise of protection and innovation. That's why this book is different it's designed to help you break through the noise, outsmart bigger competitors stay ahead of trends, and connect with your audience in authentic and meaningful ways. Through AI-enhanced creativity, newsjacking techniques, and innovative content strategies, this guide will help you stand out, engage your audience, and build long-term relationships in an industry where credibility matters.

How to Get the Most from This Book

To help you apply these strategies effectively, you'll find:

- Customizable AI prompts for every chapter to accelerate content creation
- A Quick-Start Guide for implementing AI-driven marketing strategies
- An exclusive bonus chapter with additional AI-powered growth tactics

If you're looking for ways to sharpen your marketing, outthink competitors, and harness AI effectively, this book will provide the insights and tools you need.

My Promise to You

By the end of this book, you'll have the strategies, insights, and confidence to tackle even the toughest cybersecurity and high-tech marketing challenges. This isn't about one-size-fits-all solutions - it's about practical, creative strategies that you can adapt to your business needs.

Whether you're targeting IT leaders, C-suite executives, or tech-savvy consumers, this book provides actionable insights to help you strengthen your brand, attract high-value clients, and optimize your marketing with AI-powered tools.

Quick-Start Guide

To help you implement strategies quickly, I've included:

- Customizable AI prompts for every chapter so you can create marketing content faster
- A Quick-Start Guide to cybersecurity and high-tech marketing success with step-by-step execution plans

Exclusive Bonus Chapter: AI-Powered Marketing Secrets

This book also includes an exclusive bonus chapter packed with AI-driven, ready-to-use strategies to help you optimize your marketing and sales approach.

- Generate more leads and close deals faster with AI-powered sales enablement
- Differentiate your cybersecurity or high-tech startup with precision-focused positioning strategies
- Gain a competitive edge with automation—leveraging the same AI tools used by top brands to scale marketing

These are not generic marketing strategies—they are tested, AI-optimized approaches designed to help you stay ahead in a rapidly evolving industry.

Make sure to check it out before you start implementing the strategies in this book.

The Role of AI in Modern Marketing

Just as many authors have used ghostwriters for decades (some estimates suggest up to 80% of non-fiction books involve ghostwriting), AI is now a powerful tool that enhances marketing creativity and execution. This book embraces AI as a tool, when combined with strategic thinking, helps marketers work smarter, optimize campaigns, and drive real results.

Let's Get Started

Are you ready to create marketing that's innovative, impactful, and forward-thinking? Let's begin your journey toward future proofing your marketing strategies and standing out in the cybersecurity and high-tech landscape.

Oh, and if you spot a typo or a formatting hiccup, consider it a bonus challenge - proof that even perfection isn't immune to the occasional vulnerability.

Discover More in Book Complimentary Hub:

https://911cybersecurity.com/book/

Have questions or need help? Connect with me on LinkedIn: https://www.linkedin.com/in/webdesignerny/

Table of Contents

But Wait, There's More! Bonus Chapter: The Cybersecurity & High-Tech AI Marketing Treasure Chest

Welcome to your ultimate AI Marketing Treasure Chest - a curated collection of game-changing prompts and strategies designed to revolutionize cybersecurity and high-tech marketing.

This chapter is designed to be a game-changer for cybersecurity and high-tech startups, sales teams, and marketing executives. If you're in a competitive market especially in major tech hubs like New York and California and other this is your treasure chest of high-impact, AI-driven strategies that will help you dominate your industry.

What makes this chapter different?

- *Future-Proof* – Adaptable to evolving AI trends for tomorrow's challenges.
- *Plug-and-Play* – Ready-to-use AI marketing prompts that you can deploy immediately.
- *Sales-Boosting* – Crafted to drive revenue growth and supercharge lead generation.
- *Market-Smart* – Tailored specifically for the most in-demand cybersecurity and high-tech industries.

The Reality: Tech & Cyber Threats Driving Demand

The cybersecurity and high-tech sectors in New York and California are evolving rapidly. Fueled by constant innovation, emerging security threats, and stringent regulatory demands, these industries need cutting-edge solutions and sophisticated marketing strategies to stay ahead of the competition. As digital transformations grow, so does the need for advanced technologies and adaptive marketing to effectively address both security challenges and customer demands.

In today's fast-paced digital battleground, standing still is not an option. The strategies and templates below are designed to not only address current threats but also to empower you to anticipate future challenges—turning every campaign into a competitive advantage.

Most In-Demand High-Tech & Cybersecurity Services

- *AI-Powered SaaS Solutions* – Predictive analytics, automation, and workflow optimization.
- *Zero-Trust Network Security* – Prevents unauthorized access, ensuring robust defense.
- *AI-Driven Threat Detection* – Real-time monitoring, detecting, and responding to cyber threats.
- *FinTech & Crypto Security* – Safeguarding digital transactions and securing crypto wallets.
- *MedTech & Biotech Startups* – AI-driven diagnostics, advanced analytics, and patient data security.

AI-Powered Marketing Playbook: Customizable Templates for High-Tech & Cybersecurity

1. AI-Powered Product Positioning (Customizable Template)

Customizable AI Prompt: *Generate a compelling value proposition for a [high-tech/cybersecurity] startup specializing in [specific product/service]. Focus on competitive advantages, ROI, and customer benefits.*

Example for a MedTech Startup: Generate a unique value proposition for a MedTech AI startup offering real-time patient monitoring software. Highlight its AI-driven predictive analytics, cost-saving benefits, and compliance with healthcare regulations.

11

Remember: These prompts and templates are fully customizable. Tailor them to your brand's unique voice and market needs to create campaigns that truly resonate and drive results.

2. AI-Driven Lead Generation & Demand Creation

Customizable AI Prompt: *Write a cold outreach email for a [high-tech/cybersecurity] startup targeting [ideal customer persona]. Focus on their pain points, industry challenges, and how this AI-powered solution offers the perfect fit.*

Example for AI in eCommerce Personalization:

Write a cold email targeting eCommerce CMOs about an AI personalization engine that increases customer retention and conversion rates.

3. AI-Optimized Content & Thought Leadership

Customizable AI Prompt: *Develop a blog post outline on 'How AI is Disrupting [Industry]' for [target audience]. Structure it to establish the company as a thought leader and attract high-value clients.*

Example for an AI-Powered SaaS Company:

Create a blog post outline *on 'How AI is Transforming Supply Chain Management'* targeting logistics and procurement managers.

4. AI-Powered Sales Funnel Optimization

Customizable AI Prompt: *Generate a list of persuasive sales objections and rebuttals for a [high-tech/cybersecurity] product. Ensure responses are data-driven, emphasize ROI, and alleviate hesitation.*

Example for a "CleanTech Startup":

List common objections investors might have when funding a "CleanTech startup" and provide AI-generated responses that emphasize long-term sustainability and ROI.

5. AI-Generated High-Impact Call-to-Actions (CTAs)

Customizable AI Prompt: *Write 10 high-converting CTAs for a [high-tech/cybersecurity] website landing page targeting [specific audience]. Ensure these CTAs are designed to drive action and maximize conversions.*

Example for a SaaS Free Trial Landing Page:

Start Your Free AI-Powered Workflow Automation Trial Today.

Boost Efficiency—See AI in Action with a Live Demo

Marketing within the cybersecurity and high-tech sectors is not merely about generating leads or closing sales. It's about shaping the future, building trust, and helping industries evolve to meet ever-changing challenges.

Key Takeaways:

- AI is your strategic advantage, those who embrace it early might success in their industries.
- Customization is power—these AI prompts and templates provide the flexibility to tailor your marketing strategy to your unique business needs.
- Success in high-tech marketing isn't about following trends—it's about creating them.

What to Do Next:

1. Copy the AI prompts and start experimenting today.
2. Personalize them to fit your exact business needs and target audience.
3. Launch AI-driven campaigns and begin seeing measurable results.

This isn't just a bonus chapter it's your blueprint for marketing that fuels high-growth, future-proof businesses. AI will be your co-pilot as you enhance your cybersecurity business's market position.

Customizable Text-to-Image Prompts

Tip: *Visual storytelling is key—use these image prompts to craft visuals that are as innovative and engaging as your marketing strategy.*

Below are image generation prompts you can personalize and use based on your business and marketing strategy. These can be used with AI-powered tools to create visuals that align with your marketing campaigns.

1. Product Positioning Image (AI-Powered SaaS Solutions)

Customizable AI Prompt: *Generate an image of a sleek, modern AI-powered SaaS platform interface with real-time predictive analytics displayed on a dashboard. The visual should highlight efficiency, automation, and futuristic design.*

Personalization: Add your brand colors and logo to the interface to make it look like your product's dashboard. Specify the type of service you provide (e.g., predictive analytics, workflow automation).

2. AI-Driven Lead Generation Visual (Cybersecurity Startup)

Customizable AI Prompt: *Create an image that illustrates a cybersecurity solution preventing a digital attack in real-time, with a high-tech, shield-like icon protecting sensitive data. The scene should evoke security, innovation, and trust.*

Personalization: Insert the name or tagline of your product (e.g., "Next-Gen Cyber Defense"). Choose a color palette that matches your branding and the vibe you want to project—be it corporate, futuristic, or sleek.

3. Blog Post Header Image (AI in Supply Chain)

Customizable AI Prompt: *Design an image for a blog post titled 'How AI is Transforming Supply Chain Management'. The visual should include interconnected logistics nodes, AI-powered robots, and a global map with flowing data.*

Personalization: Customize with your company's logo and branding colors. If you specialize in a specific region, add landmarks or maps specific to that area to make it more localized and relevant.

4. Sales Funnel Illustration (AI-Powered Threat Detection)

Customizable AI Prompt: *Create a visual showing a sales funnel with stages from awareness to decision, with AI-powered cybersecurity tools actively filtering and detecting threats at each stage.*

Personalization: Tailor the sales funnel to your specific product. For example, if you're focusing on crypto security, feature crypto-related icons and incorporate your product's branding at each funnel stage.

5. Customer Success Story Visual (MedTech Startup)

Customizable AI Prompt: *Generate an image of a doctor or healthcare professional using AI-powered diagnostic tools in a hospital or lab setting. Include futuristic medical technology, patient data on digital screens, and high-tech medical equipment.*

Personalization: Modify the doctor's appearance or setting to reflect your target audience (e.g., healthcare professionals in a specific market or region). You can also add your company logo and branding on the diagnostic screens for a seamless integration with your marketing materials.

6. Call-to-Action Visual (SaaS Free Trial)

Customizable AI Prompt: *Create an engaging, clean landing page visual for a SaaS free trial. Feature an eye-catching button that reads "Start Your Free Trial," with the visual showing a user-friendly dashboard with the power of AI in action.*

Personalization: Use your company's colors, logo, and product features in the dashboard. Customize the CTA button text to align with your brand's voice (e.g., "Unlock Your AI-Powered Trial Today").

How to Use These Prompts:

- Choose the relevant prompt that aligns with your current marketing initiative or campaign.
- Personalize the placeholder information (e.g., your product name, branding, target audience) in the prompts.
- Input the prompt into a text-to-image AI tool (such as DALL·E or other AI visual tools).

Your Treasure Chest Awaits: This bonus chapter is more than a collection of ideas - it's a powerful toolkit designed to transform your marketing approach. Dive in, experiment with these AI-driven prompts, and watch as your cybersecurity and high-tech campaigns become the industry's next big game changer.

Get ready to inspire awe and drive action like never before!

Chapter 1: Leveraging Cyberattacks for Strategic Marketing Success: Lessons for Cyber & High-Tech Industries.

Inspiration Before You Begin: *"The most impactful marketing campaigns don't just inform—they transform. Every cyberattack tells a story, and your solutions should be the hero."*

Marketing from the Frontlines: Lessons from Cyberattacks

Cyberattacks dominate headlines and fuel widespread concerns across industries. While these events pose significant security challenges, they also present an opportunity for cybersecurity and high-tech marketers. Strategic positioning around major breaches and security incidents can not only educate potential clients but also establish your brand as an industry leader.

In this chapter, you will learn how to turn cybersecurity incidents into high-impact marketing campaigns by leveraging real-world events, industry insights, and AI-powered content strategies.

How to Get Started with AI in Cybersecurity Marketing

For many professionals, implementing AI tools can feel overwhelming, especially if you're new to the technology. Here's a simple step-by-step guide to help you start using AI in your marketing and sales efforts:

1. **Define Your Goal**:

Example: Do you want to generate leads, improve email engagement, or create better campaigns?

Action: Write down a specific objective like: *"I want to increase email open rates by 20%."*

2. **Choose the Right Tool**:

Popular tools like **ChatGPT**, **Jasper**, and **HubSpot AI** can help with content creation, email automation, and data analysis.

Action: Start with one tool and test its capabilities.

3. **Experiment with Simple Prompts**:

Use ready-made prompts to start generating ideas.

Example Prompt: *"Create a blog post outline targeting SMB owners about the importance of proactive cybersecurity measures."*

4. **Evaluate and Adjust**:

Review the AI output and refine it to align with your brand's voice.

Action: Test different prompts to see what works best.

5. **Integrate AI with Existing Tools**:

Many AI platforms can integrate with tools like **Google Analytics** and **Mailchimp** for seamless implementation.

Action: Connect your chosen AI tool to your current workflow for optimal results.

While AI transforms cybersecurity marketing, its power extends far beyond this sector. In the high-tech industry including SaaS, hardware, and AI-driven solutions—AI marketing plays a pivotal role in demand generation, automation, and product adoption.

Expanding AI Marketing to High-Tech: A Competitive Edge

AI marketing isn't just transforming cybersecurity—it's revolutionizing software, SaaS, and high-tech industries. Whether you're launching an AI-powered SaaS platform, promoting the latest cloud security solution, or selling enterprise hardware, AI-driven marketing gives high-tech companies a unique edge.

For SaaS and Software Vendors:

- AI-driven customer segmentation: Identify user behavior and predict product adoption trends.
- AI-powered recommendation engines: Deliver personalized feature suggestions based on user actions.

For Hardware & IoT Providers:

- Predictive AI marketing: Forecast demand trends for semiconductor chips, smart devices, or AI-powered hardware.
- AI-powered customer engagement: Chatbots guide potential customers to the right product before they even reach out.

Example AI Prompt for High-Tech Marketers:

"Create an AI-driven marketing campaign for a SaaS company offering AI-powered workflow automation. The campaign should highlight efficiency improvements, competitive advantages, and an interactive demo."

Key Strategies for Leveraging Cyberattack Case Studies

1. Emphasizing Proactive Security Through Real-World Incidents

Highlighting recent cyberattacks underscores the necessity of proactive security measures. By aligning your marketing campaigns with these events, you can educate your audience on the importance of staying ahead of threats.

Example Campaign: The U.S. Treasury Hack

- **Headline:** "Could Your Business Survive a Treasury-Level Cyberattack?"
- **Message:** "The recent U.S. Treasury hack exposed critical vulnerabilities in supply chains and third-party software. Our proactive security solutions ensure such breaches are identified and mitigated before they impact your operations."
- **Call-to-Action:** "Schedule a complimentary security assessment to uncover and address potential vulnerabilities in your system."

The Business Impact of Supply Chain Attacks (Statistical Insight)

According to IBM's Cost of a Data Breach Report (2024), supply chain-related cyberattacks account for 19% of all breaches, with an average cost of $4.46 million per breach.

In the case of the U.S. Treasury hack, attackers infiltrated "Supply Chain" software breach, affecting more than 18,000 organizations worldwide, including government agencies, Fortune 500 firms, and critical infrastructure. This showcases the importance of proactive vendor risk assessments and Zero Trust strategies.

Pro Tip:

Implement continuous monitoring solutions that assess third-party risks in real time, rather than relying on annual vendor security assessments that might be outdated by the time an attack happens.

Supply Chain Attacks: Annual Growth & Financial Impact (2018-2024)

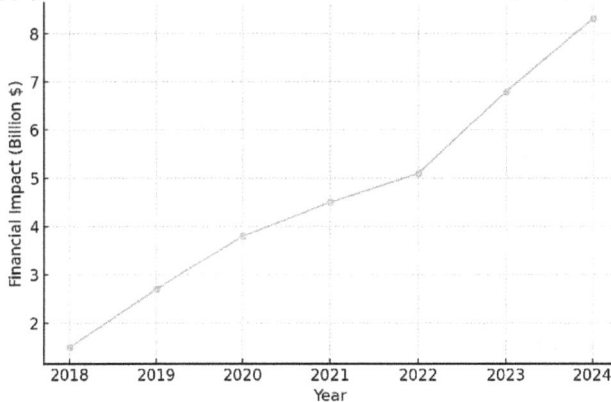

Customizable AI Prompt:

" Write a compelling LinkedIn post addressing the impact of [latest cyberattack] on SMBs. Highlight how implementing solutions such as Zero Trust, MFA, or advanced ransomware prevention can mitigate risks. Conclude with a clear call-to-action inviting readers to schedule a free security assessment or download a detailed case study."

What You Get:

A LinkedIn post tailored to SMB owners, highlighting vulnerabilities exposed by the Treasury hack, how your solutions address these issues, and a clear next step (e.g., free consultation).

2. Industry-Specific Strategies to Highlight Proactive Measures

Tailoring your approach to address the unique challenges of each industry ensures your message resonates and underscores the importance of proactive security.

For Healthcare Campaign: *"Protect Patient Data Before It's Too Late"*

Message: *"With the rise in healthcare data breaches, proactive security measures are essential to safeguard patient information and maintain trust."*

CTA: Offer a free HIPAA compliance audit to identify and address potential vulnerabilities.

Customizable AI Prompt:
"Compose a blog post for healthcare providers, discussing the increase in data breaches and the necessity of proactive security measures, with an offer for a free HIPAA compliance audit."

What You Get:
An informative blog post that educates healthcare providers on current threats and promotes your proactive security services.

Industry-Specific Marketing for High-Tech & SaaS

High-tech businesses must translate technical innovations into real-world benefits for non-technical buyers. This requires precision messaging, use-case-driven campaigns, and education-focused content.

For SaaS Companies:

- Campaign Theme: "Automate, Optimize, Scale: The Future of Work is Here"
- Messaging: Highlight how AI-driven SaaS tools reduce manual labor, improve efficiency, and optimize decision-making.
- CTA: "Start your free AI automation trial today and transform your workflow."

For AI Companies:

- Campaign Theme: "Trust in AI: Ethical, Transparent, and Powerful"
- Messaging: As AI regulations evolve, build trust and transparency in AI-powered software solutions.
- CTA: "Download our whitepaper: Building Ethical AI for the Enterprise."

For Cloud Providers:

- Campaign Theme: "99.99% Uptime is Not Enough – Why Redundancy is the Future"
- Messaging: Cloud failures cost millions—promote disaster recovery, multi-cloud solutions, and business continuity strategies.
- CTA: "Learn how multi-cloud strategies can protect your business from downtime."

Example AI Prompt for SaaS Marketers:
"Create a LinkedIn post promoting an AI-driven SaaS solution for IT teams, focusing on efficiency, automation, and reduced manual workload."

3. Crafting Emotional Narratives to Highlight the Cost of Inaction

Emotional storytelling can effectively convey the consequences of neglecting proactive security measures, making the risks more relatable to decision-makers.

Example: For Financial Services

Storyline: *"Imagine a scenario where a financial institution falls victim to a data breach due to outdated security protocols. Client trust is shattered, leading to significant financial and reputational losses. This could have been prevented with regular security assessments and proactive measures."*

Actionable CTA: Offer a free consultation to evaluate current security protocols and recommend proactive improvements.

Customizable AI Prompt:

"Develop a detailed case study template for [industry] that outlines how a company effectively responded to and recovered from a [specific cyberattack—e.g., ransomware, phishing, or supply chain breach]. Include sections for emotional storytelling, financial impact analysis, key lessons learned, and a clear call-to-action for a consultation or security audit."

What You Get:

A detailed case study that illustrates the tangible impacts of neglecting proactive security, providing a compelling narrative for potential clients.

Marketing High-Tech Failures: Lessons for Lead Generation

High-profile technology failures can be just as effective for marketing as cyberattacks. Think about cloud service outages, major AI ethics controversies, or data loss incidents. These create prime marketing opportunities to showcase preventative solutions.

Example Campaign: Learning from Major Cloud Outages

- **Headline:** "What Happens When Your Cloud Fails? A Lesson from [Cloud Provider's] Global Outage."
- **Message:** "Cloud downtime cost businesses millions in lost revenue and productivity. Our cloud redundancy solutions ensure uptime reliability for enterprises."
- **CTA:** "Schedule a risk assessment and avoid costly downtime."

Example AI Prompt for High-Tech Marketing:
"Write a case study on how a company prevented losses during a major cloud outage using redundancy solutions. Include a CTA for a consultation."

Creative Digital Campaigns and Gamification

Gamification in cybersecurity often serves as an educational tool, helping companies improve security awareness. In contrast, SaaS and high-tech brands use gamification primarily for onboarding, customer engagement, and feature adoption.

Here's how their gamification strategies differ:

Category	Cybersecurity Marketing	SaaS & High-Tech Marketing
Primary Goal	Security awareness & behavior change	Customer engagement & feature adoption
Core Audience	Employees, IT teams, decision-makers	Users, trial customers, product adopters
Typical Format	Quizzes, phishing simulations, security challenges	Interactive tutorials, onboarding milestones, gamified rewards
Psychological Hook	Fear-based urgency (threat prevention)	Curiosity & reward-driven motivation
Example Campaign	"Spot the Phishing Attack" Quiz (employees earn security certification badges)	"Complete Onboarding & Unlock a Premium Feature" (users gain VIP perks)
Engagement Metric	Risk awareness improvement, incident response speed	Product activation, retention, and upsell conversions
Incentives Used	Digital certificates, leaderboard rankings, company-wide recognition	Free trials, exclusive discounts, access to advanced features

To truly "WOW" your audience, move beyond traditional campaigns and incorporate gamification and interactive elements into your marketing strategy.

Example: Gamified Cybersecurity Quiz

Concept: *"How Safe Is Your Business?"*

Execution: Use platforms like Typeform or Kahoot to design a short quiz assessing cybersecurity readiness.

Offer a free consultation or downloadable guide as a reward for completing the quiz.

Example Question: "How often does your team update passwords?"

- Every month
- Every year
- Only after a breach

Customizable AI Prompt: "Create an interactive quiz for [industry] companies to assess their cybersecurity readiness. Include questions on password management, employee training, and backup systems."

Example: SaaS Free Trial Gamification

Concept: "Complete the onboarding challenges & unlock a free 30-day upgrade!"

Execution: Assign user tasks such as completing a tutorial, setting up an automation, or integrating a tool.

Reward: A premium feature unlock for users who complete the onboarding journey.

Example AI Prompt for Gamification:
"Design a gamification strategy for a SaaS product trial, including engagement incentives and milestone rewards."

Example: Interactive Webinar Series

Concept: Host a webinar titled ***"Defend Against Cyberattacks in Real Time"*** with live simulations of phishing attempts.

Execution: Allow attendees to make real-time decisions during a simulated attack, demonstrating the importance of proactive security measures.

Reward participation with exclusive access to whitepapers or eBooks.

Customizable AI Prompt: *"Draft a webinar outline for a live simulation of a phishing attack, including key takeaways and a promotional offer for attendees."*

Gamifying High-Tech & SaaS Marketing

Gamification doesn't just make cybersecurity engaging—it supercharges high-tech customer acquisition. Tech adoption barriers are high, so gamification simplifies onboarding, product training, and customer retention.

Example: SaaS Free Trial Gamification

- Concept: "Complete the onboarding challenges & unlock a free 30-day upgrade!"
- Execution: Assign user tasks such as completing a tutorial, setting up an automation, or integrating a tool.
- Reward: A premium feature unlock for users who complete the onboarding journey.

Example: AI Chatbot Gamification

- Concept: "Test Our AI Assistant – Can You Outperform It?"
- Execution: Invite users to challenge an AI chatbot in data sorting, scheduling, or email responses.
- Outcome: Users engage with the product in a fun, competitive environment while learning its real-world applications.

Example AI Prompt for Gamification:
"Design a gamification strategy for a SaaS product trial, including engagement incentives and milestone rewards."

Quick-Start: AI-Driven Campaign Framework

1. **Objective**: Promote proactive security measures to prevent cyberattacks.
2. **Target Audience**: Decision-makers in [specific industry].
3. **Messaging**:
   "Don't wait for a breach to take action. Our proactive security solutions identify and mitigate risks before they become threats."
4. **AI Tools**: Utilize AI-driven analytics to identify industry-specific vulnerabilities and tailor messaging accordingly.
5. **Metrics to Track**: Engagement rates, number of assessments scheduled, and conversion rates post-assessment.

Customizable AI Prompt: *"Develop a comprehensive marketing campaign targeting [industry] decision-makers, emphasizing the importance of proactive security measures and offering a free vulnerability assessment."*

What You Get:
A detailed marketing campaign outline that highlights the urgency of proactive security and encourages potential clients to engage with your services.

Exclusive Resources for Marketing Success

https://911cybersecurity.com/book/

Chapter 2: Newsjacking & Industry Trends: A Marketer's Guide to Cybersecurity & High Tech

Newsjacking for Impact: Turning Headlines into High-Converting Campaigns

Inspiration Before You Begin: *"The best marketing campaigns don't just inform—they inspire. Think like your customer. How would a cybersecurity breach feel to them? Fear, urgency, and trust are your keys to connection."*

Newsjacking is the practice of leveraging breaking news or trending industry events to promote your brand, engage your audience, and position your company as an authority. In cybersecurity and high-tech industries, where news about breaches, regulations, and technological advancements moves rapidly, newsjacking is a critical marketing tool.

In this chapter, we'll explore how cybersecurity and high-tech companies can capitalize on industry trends and news to create high-converting campaigns.

Why Newsjacking Works for Cybersecurity and High-Tech Marketing

- **Immediate Relevance:** Companies that respond quickly to breaking cybersecurity news can ride the wave of heightened interest.
- **Trust Building:** Thoughtful analysis of a security breach or industry shift positions your brand as an authority.
- **Lead Generation:** Offering immediate insights and solutions attracts potential clients who are looking for expert guidance.

- **SEO and Social Media Visibility:** News-related content can rank higher in search results and gain significant traction on LinkedIn, Twitter, and other platforms.

Key Strategies for Successful Newsjacking

1. Monitor Industry News Constantly

- Set up Google Alerts for cybersecurity topics and emerging threats.
- Follow leading cybersecurity and tech news sources such as Wired, Dark Reading, and TechCrunch.
- Use AI-driven sentiment analysis tools to track trending discussions on social media.

2. Act Fast—Timing is Everything

- Newsjacking works best when you react within the first **24-48 hours** of an event.
- Assign a dedicated response team to draft and approve content quickly.

3. Create a Variety of Content Formats

- Write an expert **LinkedIn post** analyzing the situation.
- Publish a **blog article** with security best practices related to the event.
- Host a **live webinar** discussing implications and offering solutions.
- Create **short-form videos** summarizing key insights for social media.

4. Ensure Your Angle Adds Value

- Instead of merely reporting the news, provide analysis and actionable takeaways.

- Example: If a major cloud provider experiences an outage, a cybersecurity firm could publish: "5 Steps to Ensure Business Continuity Amid Cloud Disruptions."

5. Integrate AI for Faster Response

- Use AI-driven content generation tools to quickly draft newsjacking posts.
- Leverage predictive analytics to foresee potential trends before they go viral.

Example AI Prompt: *"Write a LinkedIn post targeting SMB owners about lessons learned from the recent U.S. Treasury hack. Emphasize the importance of Zero Trust architecture and end with a call-to-action for a free security assessment."*

Visual Aid: Create a timeline of recent high-profile cyberattacks, showing how they could have been mitigated with the right solutions.

Strategic Foresight: Lessons from The Art of War

"In the midst of chaos, there is also opportunity."

This principle from The Art of War underscores why newsjacking is so effective. The rapid rise of cyberattacks creates moments of uncertainty and fear, which your marketing can address with timely, relevant campaigns. By anticipating potential threats and preparing campaigns in advance, you position your brand as a trusted leader before competitors can react.

Example Application:
When the U.S. Treasury breach dominated headlines, a

cybersecurity firm launched a proactive campaign within 24 hours. Their blog post, "How to Protect Your Supply Chain from the Next U.S. Treasury HACK," included:

- A checklist for assessing third-party vendors.
- A free consultation offer for supply chain security audits.

Results:

- Website Traffic: 300% increase in 3 days.
- Lead Conversions: 20% jump in consultation sign-ups.

Case Study: Newsjacking in Action

A managed service provider (MSP) capitalized on a high-profile ransomware attack targeting SMBs. Within 48 hours, their team launched a multi-channel campaign titled "Ransomware Readiness: How to Protect Your Business Now". The campaign included:

1. A Blog Post: Outlining the common vulnerabilities exploited by ransomware.
2. A Webinar: Featuring live Q&A with cybersecurity experts.
3. Email Outreach: Offering free ransomware vulnerability assessments.

Results:

- Email Click-Through Rate: 15% (double the industry average).
- Webinar Attendance: 400+ participants.
- New Business Contracts: $150,000 in sales directly linked to the campaign.

Takeaway: Speed and relevance are critical. By responding quickly to breaking news, you can engage prospects at the peak of their interest.

Cyberattack Timeline for Newsjacking & Marketing Opportunities

1. Massive Data Breach at a Global Financial Institution

Attack: Sensitive financial data of millions of customers was exposed due to cloud misconfiguration.

Newsjacking Opportunity: Cloud security providers highlighted the dangers of misconfigured settings and poor IAM (Identity & Access Management) policies.

Marketing Tactics: Targeted LinkedIn ads, webinars on "Preventing Cloud Data Leaks," and email campaigns with free Cloud Security Checkups.

AI Prompt for Marketers: *"Create a LinkedIn post targeting CFOs and IT leaders about preventing cloud misconfigurations that lead to data breaches. Offer a free cloud security audit as a CTA."*

2. Nation-State Supply Chain Attack

Attack: Hackers infiltrated a widely used software provider, injecting malware that spread across thousands of businesses.

Newsjacking Opportunity: Zero Trust security firms promoted solutions for "Never Trust, Always Verify" architectures.

Marketing Tactics: PR articles, Twitter threads explaining supply chain risks, and lead magnet whitepapers on Zero Trust best practices.

AI Prompt for Marketers: *"Generate an email campaign for cybersecurity vendors addressing supply chain security risks. Highlight a real-world attack and offer a guide on Zero Trust implementation."*

3. Ransomware Attack on Critical Infrastructure

Attack: A ransomware attack shut down fuel pipelines, causing economic disruption and widespread panic.

Newsjacking Opportunity: Endpoint protection companies promoted anti-ransomware solutions.

Marketing Tactics: Geo-targeted ads, urgent email campaigns, and incident response playbooks for potential victims.

AI Prompt for Marketers: *"Write a blog post for energy and utility companies on preventing ransomware attacks. Include a checklist for securing OT/ICS environments and a call-to-action for a security risk assessment."*

4. Insider Threat: Disgruntled Employee Leaks Data

Attack: A former employee with admin access leaked sensitive company data before leaving.

Newsjacking Opportunity: Insider threat monitoring firms showcased privileged access control solutions and employee behavior analytics.

Marketing Tactics: Webinar on "Preventing Insider Threats," gated research reports, and risk scoring calculators.

AI Prompt for Marketers: *"Create a social media thread for HR and IT teams about the risks of insider threats. Offer a free security checklist for monitoring employee access."*

5. AI-Powered Phishing Campaign Targets Executives

Attack: A deepfake-powered phishing scam tricked executives into wiring millions to attackers.

Newsjacking Opportunity: Email security & AI detection companies promoted real-time phishing defense solutions.

Marketing Tactics: Thought leadership articles, live attack simulations, and CISO-targeted ads.

AI Prompt for Marketers: *"Draft a press release on AI-powered phishing attacks, positioning your company's email security product as the best defense. Include industry statistics and a customer success story."*

Key Takeaways for Effective Newsjacking

- React Fast: Publish content within 24-48 hours of an incident.
- Educate, Don't Exploit: Offer actionable insights, not fear-mongering.
- Segment Audiences: Use geo-targeting or industry-specific messaging.
- Leverage AI & Automation: Use AI-powered tools to quickly generate campaigns.

Emotional Storytelling: Engaging Decision-Makers

Cybersecurity is not just a technical challenge; it's a human one. By framing your messaging around the emotional impact of cyberattacks - loss of trust, reputational damage, financial hardship - you can connect with decision-makers on a deeper level.

Example: *"Imagine waking up to find your customer data compromised, your reputation in tatters, and your business grinding to a halt. Our solutions ensure that this nightmare remains a fiction, not a reality."*

Visual Element: Add an infographic comparing the cost of proactive cybersecurity measures versus recovering from a data breach.

The Role of Data and Statistics in Building Credibility

Credibility is the cornerstone of successful cybersecurity marketing. Decision-makers need to trust that your solutions are effective. Supporting your claims with data—such as success rates, case studies, and industry benchmarks—can make all the difference.

Actionable Tip: Incorporate visual aids like infographics and charts to present data compellingly.

Clever Fact: *"On average, organizations take 287 days to identify and contain a data breach. Imagine how much damage could be avoided with proactive measures."*

AI-Driven Marketing Benefits:

1. Increase in Engagement Rates with Personalization: +37%
2. Improved Lead Qualification through Predictive Analytics: +45%
3. Higher Conversion Rates via AI Chatbots: +32%
4. Reduction in Campaign Costs using AI Optimization: +28%

Discover More: Exclusive Resources

https://911cybersecurity.com/book/

Chapter 3: Mastering AI for Cybersecurity & High-Tech Marketing

Inspiration Before You Begin: *"AI doesn't replace creativity, it amplifies it. Think of AI as your assistant, crunching numbers and generating ideas, so you can focus on strategy and storytelling."*

AI is transforming cybersecurity and high-tech marketing, making it smarter, faster, and more precise than ever before.. Today's marketers no longer rely on guesswork— **AI-powered tools provide real-time insights**, automate engagement, and optimize content strategies to **reach the right audience at the right time**.

In this chapter, we'll explore **AI-driven marketing strategies** that enhance personalization, improve efficiency, and future-proof your cybersecurity brand.

Why AI is a Game-Changer for Cybersecurity and High-Tech Marketing

- Advanced Data Analysis: AI can process large-scale security trends and customer data in real-time, identifying threats and opportunities faster than traditional methods.
- Hyper-Personalization: AI-powered marketing platforms tailor messages and campaigns to specific audience segments, increasing engagement rates.
- Predictive Marketing: AI can anticipate emerging cybersecurity threats, allowing brands to create proactive content and solutions ahead of industry competitors.
- Automation & Efficiency: AI streamlines content creation, email marketing, and customer interactions, reducing manual work and improving response times.

Personalization at Scale

AI allows marketers to deliver highly personalized content to prospects at scale. By analyzing user behavior, preferences, and past interactions, AI-powered tools can help you craft messages that resonate on an individual level.

Case Study: AI-Driven Personalization in Cybersecurity SaaS Marketing

Company: CyZTech Solutions **(Fictional)**

Challenge: Low engagement rates on email campaigns. Struggled to convert free trial users into paying customers.

AI Solution: The company used predictive AI analytics to segment prospects based on industry and security needs. It then implemented an AI-driven email campaign, dynamically inserting customized product recommendations.

Results:

- 30% increase in email open rates with AI-generated subject line testing
- 18% boost in free trial conversions
- 40% decrease in customer churn due to improved onboarding

Key Benefits:

Build Trust Through Hyper-Relevant Messaging

Practical Advice: Use data analytics tools like HubSpot to segment your audience by behavior and preferences. For example, if a lead downloads an eBook on endpoint protection, send a follow-up email with additional resources and a demo offer.

Creative Thought: Frame your messaging around the customer's specific concerns. For instance, highlight

"Peace of Mind for Your Remote Workforce" as a tagline for endpoint security services.

1. **Increase Engagement Rates with Tailored Campaigns**

Practical Advice: Launch industry-specific campaigns (e.g., "Protecting Retailers from POS Attacks" for retail or "HIPAA-Compliant Solutions for Healthcare"). Use storytelling to make technical solutions relatable.

Creative Thought: Develop a campaign that compares proactive measures to real-world consequences. For instance, "One Phishing Email Can Cost Millions—Be Proactive."

2. **Shorten Sales Cycles by Addressing Pain Points Directly**

Practical Advice: Include case studies in your outreach to showcase proven results. For example, "How We Reduced Downtime by 85% for a Manufacturing Client."

Creative Thought: Use urgency-driven messaging. *"Did You Know? Cyberattacks Increase 35% During the Holiday Season. Ensure You're Prepared Today."*

Customizable AI Prompt with Placeholders: *"Create a [type: email, social media post, blog] for [audience: SMB owners, IT managers] in [industry: healthcare, finance] located in [geographic area: Los Angeles, Texas]. Highlight [specific pain point: outdated security, phishing risk] and end with [call-to-action: free consultation, demo]."*

Impact of AI-Driven Personalization on Lead Conversions

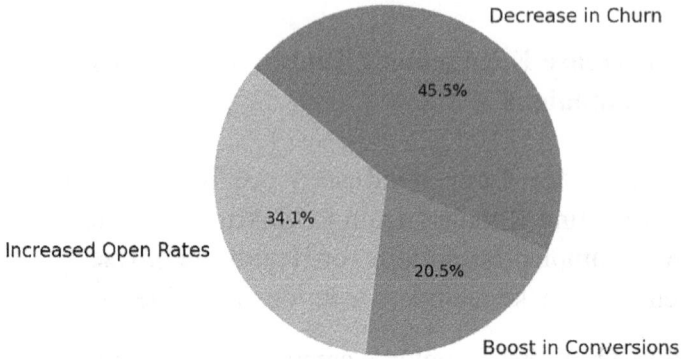

Decrease in Churn

45.5%

34.1%

Increased Open Rates

20.5%

Boost in Conversions

Why Customization Matters:

- Targeting specific audiences ensures messages resonate.
- Geographic customization allows you to address local market demands and regulations.
- AI tools can analyze regional data to personalize content further.

Visual Aid: Include an infographic comparing engagement rates for generic versus personalized campaigns.

AI-Driven Personalization for High-Tech Marketing

While cybersecurity marketing uses AI to build trust and security awareness, high-tech industries leverage AI personalization to enhance customer experiences, improve product adoption, and increase engagement.

For SaaS & Cloud Companies:

- AI-driven segmentation identifies churn risk and triggers automated retention campaigns.
- Example: If a SaaS user stops using key features, an AI system automatically emails a tutorial or offers support.

For AI & Machine Learning Vendors:

- AI-powered personalization recommends software features based on past usage.
- Example: AI tools in photo editing software suggest new AI filters based on a user's editing style.

For High-Tech Hardware Companies:

- AI tracks customer buying behavior to upsell compatible products.
- Example: A cloud server company suggests storage upgrades based on usage patterns.

Example AI Prompt for High-Tech Marketers:
"Generate a personalized email sequence for a SaaS company that detects customer inactivity and re-engages them with feature tutorials and a discount offer."

Advanced AI Prompt Engineering for Cybersecurity Marketing

AI prompt engineering enables marketers to customize campaigns for specific audiences and needs, transforming generic ideas into tailored strategies. Here are some examples:

Personalized Campaigns:

- Prompt Example:
  "Write an email targeting SMB owners about the importance of endpoint protection. Highlight common threats like ransomware and offer a free consultation."
- Outcome: Tailored messages that resonate with SMB pain points, increasing engagement and conversions.

Audience Segmentation:

- Prompt Example:
  "Generate three LinkedIn post variations targeting healthcare IT managers about securing patient data. Include statistics and a call-to-action for a demo."

- Outcome: Posts optimized for specific industries and decision-makers.

Trend Forecasting:

- Prompt Example: *"Analyze the top cybersecurity threats for Q1 and recommend marketing strategies to address them."*
- Outcome: Data-driven insights that inform proactive campaigns.

AI Personalization for High-Tech vs. Cybersecurity While AI-powered personalization is widely used in cybersecurity marketing, its potential extends into high-tech industries like:

B2B SaaS Marketing: AI-driven chatbots that analyze real-time usage data and suggest tailored product features.

Cloud Solutions Providers: AI-powered emails that predict pain points before customers experience them.

Example Prompt for Marketers:
"Write a LinkedIn post for a cloud storage company, warning users about potential AI storage misconfigurations and how to prevent them."

Predictive Analytics for Campaign Optimization

Predictive analytics uses historical data to forecast future trends, enabling marketers to make data-driven decisions. Whether it's identifying the best time to launch a campaign or predicting which leads are most likely to convert, predictive analytics can significantly enhance your ROI.

Case Example: A small MSP used predictive analytics to identify that their healthcare clients were most likely to purchase ransomware protection in Q1, enabling them to time their campaigns perfectly.

Customizable AI Prompt Template: *"Analyze historical sales data for [region: Northeast US] to identify the best time to launch a campaign targeting [industry: healthcare, retail]. Include seasonal trends and engagement patterns."*

Actionable Tip: Use A/B testing to validate predictions and refine your strategies.

AI-Powered Predictive Analytics in High-Tech Marketing

Predictive analytics allows high-tech companies to forecast customer needs, optimize pricing, and anticipate churn risks.

For SaaS Companies:

- AI analyzes historical customer data to predict which users will churn before it happens.
- Example: If a user stops engaging with a SaaS platform for 30 days, predictive AI triggers a personalized retention campaign.

For Hardware & IoT Vendors:

- Predictive analytics forecasts product demand and adjusts marketing campaigns before trends shift.
- Example: A semiconductor company uses AI to predict a spike in demand for GPU chips and adjusts marketing efforts accordingly.

For AI & Cloud Computing Firms:

- AI-driven insights identify which enterprise clients need upgrades based on real-time infrastructure usage.

- Example: A cloud provider detects high CPU usage trends and proactively offers performance-optimized upgrades.

Example AI Prompt for High-Tech Marketers:

"Create an AI-driven marketing strategy for a cloud provider that predicts customer demand spikes and automatically adjusts email promotions to upsell higher-tier plans."

Integrating AI into Your Marketing Workflow

AI tools can streamline your marketing processes, saving time and boosting efficiency. Here's how to get started:

Choose the Right Tools:

- Use platforms like Jasper or ChatGPT for content generation.
- Leverage predictive analytics tools (e.g., Salesforce Einstein, HubSpot AI) to identify trends and optimize timing.

Automate Routine Tasks:

- Deploy chatbots for 24/7 lead qualification.
- Automate follow-up emails based on user behavior (e.g., abandoned downloads).

Test and Iterate:

- Use AI-powered A/B testing to refine messaging and visuals. For example, test two versions of a LinkedIn ad targeting financial institutions to determine which generates more leads.

Practical Tip: Begin with one AI tool to avoid overwhelming your team. Gradually integrate more advanced solutions as your comfort level grows.

AI Chatbots for Lead Generation

AI chatbots are the 24/7 sales assistants your business needs. They answer customer questions, qualify leads, and guide prospects through the sales funnel all in real-time.

Why this matters: By automating routine inquiries, your team can focus on closing deals and high-value customer interactions.

Advanced Customization Features:

- Location-Based Insights: Adjust chatbot responses based on visitor location (e.g., "Local regulations in California require...").
- Industry-Specific Questions: Tailor flows for industries like healthcare (e.g., "Are you HIPAA compliant?") or retail (e.g., "How secure is your POS system?").

Customizable AI Prompt Template: *"Create an AI chatbot script for a cybersecurity website that initiates qualification by asking targeted, industry-specific questions (e.g., for retail managers or healthcare IT leaders). The script should guide visitors through a series of questions to assess their needs, provide tailored resource recommendations, and seamlessly offer an option to schedule a consultation."*

AI Chatbots vs. Human Sales Teams: Response Speed & Conversion Rates

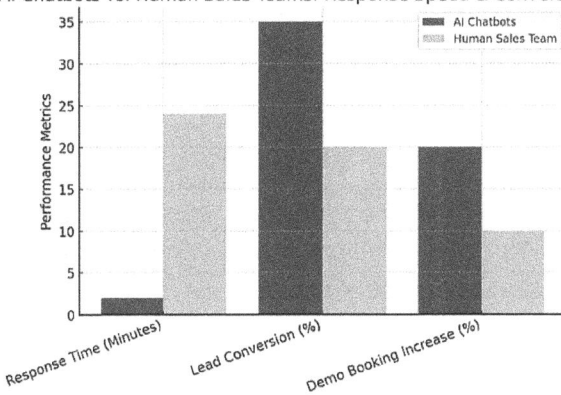

45

AI Chatbots for SaaS, Cloud, and High-Tech Lead Generation

While cybersecurity chatbots handle security risk concerns, high-tech companies use chatbots to guide users through complex software and hardware solutions.

For SaaS Companies:

- AI chatbots act as automated onboarding assistants.
- Example: A CRM SaaS tool deploys a chatbot that walks users through the setup process step-by-step.

For AI & Automation Software:

- AI-powered chatbots help users discover relevant features.
- Example: An AI design software suggests advanced tools based on past project history.

For Cloud Providers & Hardware Vendors:

- Chatbots analyze customer queries and recommend the right hardware setup.
- Example: A chatbot for a cloud server company helps users select the best hosting plan based on workload predictions.

Example AI Prompt for High-Tech Marketers:

"Create a chatbot script for a SaaS platform that helps new users onboard by recommending tutorials, feature walkthroughs, and support FAQs based on their role and usage patterns."

Fact-Checked Stats for Sales Enablement:

- 60% of SMBs close within 6 months of experiencing a cyberattack. (Source: National Cyber Security Alliance)
- 88% of buyers are more likely to purchase from a salesperson who acts as a trusted advisor rather than a traditional seller. (Source: LinkedIn Sales Report)

- Companies using tailored sales scripts and materials see 33% higher conversion rates than those using generic pitches. (Source: HubSpot)

Clever Facts:

- A great pitch is 10% words and 90% timing. Hit the right nerve, and the sale is yours.
- In cybersecurity, trust isn't just built it's encrypted.
- Cyberattacks don't wait for the fiscal quarter to end. Your sales team shouldn't either.

AI-Powered vs. Traditional Marketing Strategies

- Chart Title: "The Impact of AI on Marketing Efficiency"
- Suggested Data: Lead Generation Speed: AI-powered (3x faster).
  Engagement Rates: AI (40% higher with personalized emails).
- Campaign ROI: AI campaigns show a 25% improvement over traditional methods.

Key Takeaways & Next Steps

- AI enables cybersecurity marketers to optimize engagement, automate tasks, and predict emerging trends.
- AI-driven content generation and personalization significantly improve conversion rates.
- Predictive analytics help marketers stay ahead of industry trends and customer needs.
- Businesses should continually refine AI strategies based on performance data to maximize effectiveness.

Future-Proof AI Marketing Insight: The Next Big Shift

The next wave of AI marketing in cybersecurity and high-tech will **move beyond automation to anticipation.** The most successful brands won't just react to customer behavior, they'll predict and shape it.

- **The Next Big AI Evolution: AI-Powered Storytelling**
  Most AI-generated content today is focused on automation and efficiency. However, the next wave of AI marketing will be about storytelling that mimics human creativity.
- AI will learn from past successful campaigns and automatically create compelling narratives tailored to cybersecurity audiences.
- Future AI tools will generate sales pitches, landing pages, and email sequences that feel fully human—but with the precision of AI-driven insights.

Actionable Strategy for Today:
To stay ahead, start experimenting with AI tools that **create interactive content** such as AI-powered video explainers or chatbots that **generate dynamic stories** for cybersecurity education.

Exclusive Resources for Marketing Success

https://911cybersecurity.com/book/

Chapter 4: AI-Powered Sales Enablement: Unlocking Growth in Cyber & High Tech

Sales Supercharged: AI Tools and Insights for High-Tech Teams

Inspiration Before You Begin: *"Empowered sales teams don't just close deals—they build trust. Equip your team with the knowledge and tools to become advisors, not just sellers."*

Comprehensive Training on Cybersecurity Trends

Staying ahead of the curve in cybersecurity requires continuous learning. Sales teams must understand not only the solutions they're selling but also the evolving threat landscape.

Key Benefits:

1. Build Confidence When Addressing Technical Questions

Practical Advice: Host monthly sessions focused on explaining cutting-edge threats like AI-driven phishing attacks or supply chain vulnerabilities.

Creative Thought: Use gamification to test knowledge post-training with quizzes or team challenges. Reward top performers with incentives.

2. Position Your Team as Trusted Advisors

Practical Advice: Teach teams how to present solutions through storytelling. For example, explain how a company prevented a breach with your endpoint protection solution.

Creative Thought: Develop industry-specific "What if?" scenarios that show prospects the tangible value of your solutions.

3. Improve Credibility with Decision-Makers

Practical Advice: Share industry statistics during pitches, such as "60% of SMBs close within 6 months of a cyberattack."

Creative Thought: Equip your team with case studies that demonstrate measurable results, such as "Our solution reduced ransomware risk by 70% for a healthcare provider."

Chart Suggestion: A bar chart comparing the performance of trained versus untrained sales teams in closing deals.

Bridging the Gap: Aligning Marketing and Sales

Marketing and sales alignment is one of the most powerful strategies for increasing revenue and improving customer experiences. Yet, studies show that 60% of sales teams feel disconnected from marketing efforts, leading to missed opportunities and fragmented messaging.

Key Strategies to Align Teams:

1. **Shared Dashboards for Sales & Marketing Alignment** – Use CRM tools like Salesforce, HubSpot, or Zoho to create real-time lead tracking dashboards.

   These dashboards should show:

- Top-converting content from marketing campaigns.
- Sales team follow-up actions on leads.
- Engagement metrics to prioritize high-intent prospects

2. **Feedback Loops:**

Establish bi-weekly meetings where sales teams share insights on customer objections and marketing adapts messaging accordingly.

3. Unified Messaging:

Develop a shared style guide to ensure consistent messaging across all touchpoints.

Practical Exercise:
Host a "Sales-Marketing Hackathon" where both teams collaborate to design a campaign targeting a specific industry. For example, create a joint campaign for SMB healthcare providers focused on ransomware protection.

Tailored Sales Materials and Scripts

Generic pitches don't work in cybersecurity. Tailored sales materials and scripts ensure that your team addresses the unique needs of different industries and pain points.

Key Strategies:

1. Create Industry-Specific Content

Develop "Top 5 Tips" guides for sectors like healthcare, retail, and education. For example, "5 Cybersecurity Must-Haves for Retailers During the Holiday Season."

2. Highlight ROI with Metrics

Share visuals comparing the cost of a data breach with the cost of implementing proactive measures.

Example Sales Script: *"Hi [Prospect Name], I'm reaching out because we've helped organizations in [industry]*

reduce their risk of [specific threat, e.g., phishing attacks]
by [specific benefit, e.g., 90%]. Can we schedule a quick
call to discuss how we can do the same for your team?"

Customizable AI Prompt: *"Write a sales script targeting*
[industry: finance, healthcare] about [solution: endpoint
protection, ransomware mitigation]. Include statistics and
a strong call-to-action."

Customizable AI Prompt: *"Draft a [type: training*
module, sales script, email template] for [team: sales,
marketing] targeting [audience: SMBs, IT managers].
Focus on [threat: ransomware, phishing] and include
[interactive elements: case studies, role-playing
exercises]."

Empowering Sales Teams with AI

AI tools can supercharge sales enablement by providing
actionable insights, automating repetitive tasks, and
personalizing outreach. Here's how to integrate AI into
your sales workflow:

1. **Predictive Lead Scoring:**

AI can analyze data to rank leads based on their likelihood
to convert, helping sales prioritize efforts.

Example Tool: HubSpot's AI-powered lead scoring
feature.

2. **Automated Follow-Ups:** Use AI to send
    personalized follow-ups triggered by prospect
    behavior, such as downloading a whitepaper or
    attending a webinar.
3. **Sales Coaching:**

AI platforms like Gong can analyze sales calls and provide feedback on tone, pacing, and key objections to address.

AI-Powered Personalization & Outreach Automation

Customizable AI Prompt for Analyzing Prospect Behavior in Real-Time: *"Analyze user engagement data for [target audience: IT managers, CISOs] in [industry: healthcare, finance]. Identify patterns in interactions with [content: webinars, product demos] and suggest the best timing for outreach."*

Customizable AI Prompt for Personalizing Outreach Messages: *"Generate an email sequence tailored to [prospect type: SMB owner, enterprise IT leader] based on their past interactions with [previous content: case studies, webinars]. Include personalized recommendations and a strong call-to-action."*

Customizable AI Prompt for Automating Scheduling & Follow-Ups: *"Create a chatbot script that schedules follow-ups and meetings based on prospect engagement with [website section: pricing page, demo request]. Ensure responses are customized based on previous interactions."*

AI Chatbots in Cybersecurity Sales: Do They Work?

New Data from Gartner:

- AI-powered sales chatbots increase lead qualification efficiency by 45%.
- 40% of cybersecurity companies using chatbots saw a 20% increase in customer engagement.
- Chatbots reduce the average response time from 24 hours to just 2 minutes, boosting lead conversion rates.

Example: AI-Powered Chatbot for a Security Vendor

A cybersecurity SaaS provider implemented an AI chatbot on their website that:

- Instantly answered product-related queries
- Scheduled live demos automatically
- Analyzed visitor behavior to personalize recommendations

Results:

- 20% increase in demo bookings
- 35% higher visitor-to-lead conversion rates
- 50% reduction in human-assisted chat support costs

AI Chatbots Vs Human Sales

	Metric	AI Chatbots	Human Sales Team
1	Response Time (Minutes)	2	24
2	Lead Conversion (%)	35	20
3	Demo Booking Increase (%)	20	10

Fostering Collaboration Between Marketing and Sales Teams

Alignment between marketing and sales is crucial for converting leads into customers. Shared goals and open communication ensure a seamless buyer journey.

Key Benefits:

1. Improves Lead Handoff and Follow-Up

Practical Advice: Use CRM systems to create shared dashboards for marketing and sales teams to track lead progress.

Creative Thought: Host weekly alignment meetings where teams exchange feedback on campaign results and lead quality.

2. Provides Consistent Messaging Across Touchpoints

Practical Advice: Create shared messaging templates that sales and marketing teams can customize.

Creative Thought: Develop an internal style guide for pitches and outreach materials.

3. Enhances Understanding of Buyer Personas

Practical Advice: Co-create buyer personas by combining sales insights with marketing analytics.

Creative Thought: Run a "Persona Workshop" to ensure both teams align on key audience traits.

Customizable AI Prompt Template: *"Draft an email from marketing to sales summarizing the top-performing content and how it aligns with current lead behavior."*

Future-Proof Sales Enablement Tip: The Next Big Shift

AI-Driven Virtual Sales Assistants Will Replace Cold Calling Traditional sales calls are losing effectiveness. Future-forward cybersecurity and high-tech sales teams are adopting AI-powered virtual sales assistants that:

- Analyze prospect behavior in real time before outreach.
- Personalize outreach messages based on past interactions.
- Automate scheduling and follow-ups, ensuring no lead is left behind.

Actionable Strategy for Today:

Start testing AI-powered virtual assistants that can engage prospects and schedule sales calls without manual intervention. These tools can increase efficiency while ensuring prospects receive a tailored approach.

Test AI-powered virtual assistants like Drift, Conversica, or HubSpot's AI assistant. These tools:

- Qualify leads automatically based on engagement.
- Deliver real-time product recommendations based on website activity.
- Schedule calls instantly without human intervention, increasing conversions by up to 35%.

AI-Driven Sales Enablement vs. Traditional Sales Tactics

Why It Matters: This comparison highlights how AI sales automation improves efficiency, personalization, and conversions over traditional methods.

Sales Factor	Traditional Sales Approach	AI-Powered Sales Approach	Impact
Lead Qualification	Manual sorting by sales team	AI predictive lead scoring based on behavior data	Higher accuracy, prioritizing high-intent leads
Sales Outreach	Cold emails with generic messaging	AI-personalized outreach adjusting in real time	Increased response and conversion rates
Follow-Up Timing	Manual follow-ups at fixed intervals	AI determines the best time based on engagement	Maximized engagement & minimized drop-off
Objection Handling	Standard scripts & manual adaptation	AI-driven adaptive scripts based on customer response	More persuasive & targeted responses
Lead Nurturing	Pre-set email sequences	AI-driven behavior-triggered email sequences	Higher email open & response rates

AI-Powered Predictive Analytics for Lead Prioritization

How It Works: AI platforms like 6sense & Demandbase predict which leads are most likely to convert based on online behavior, company size, and cybersecurity needs. Example: If a company searches for Zero Trust security on multiple cybersecurity blogs, AI automatically triggers a lead nurturing campaign for targeted outreach.

Customizable AI Prompt: *"Generate an AI-driven lead prioritization model that scores prospects based on website*

56

behavior, whitepaper downloads, and webinar engagement."

Supercharge AI-Driven Sales Enablement

AI-Powered Lead Scoring: Prioritize the Right Prospects. Not all leads are equal—AI helps sales teams focus on high-converting buyers.

Hyper-Personalized AI Outreach: Tailor Messaging at Scale. AI eliminates generic outreach, ensuring personalized engagement across all touchpoints.

AI Sales Coaching: Train Teams in Real-Time. AI-powered coaching transforms sales reps into elite closers.

AI-Generated Smart Follow-Ups: Automate Lead Nurturing. 80% of sales require 5+ follow-ups—AI ensures you never lose a deal.

AI-Powered Sales Content: Deliver the Right Materials Instantly. AI ensures sales teams send the best case studies, whitepapers, and ROI reports at the right time.

Chapter 5: Becoming an Industry Authority: Leadership Strategies for Cyber & High Tech

Establishing Industry Leadership in Cybersecurity and High-Tech

Inspiration Before You Begin: "Thought leadership isn't just about being seen - it's about earning trust and driving change. Establish yourself as the go-to expert, and your customers won't just seek you out - they'll rely on you."

Content Marketing Strategies

Creating high-value content that educates and inspires your audience is key to building thought leadership. Your content should address your audience's pain points, offer actionable solutions, and position your brand as a trusted resource.

Key Strategies:

Develop Evergreen Content:

- Focus on topics that remain relevant, such as "The Ultimate Guide to Zero Trust" or "5 Steps to Prevent Ransomware Attacks." Instead of just listing steps, provide specific examples. For instance, in "5 Steps to Prevent Ransomware Attacks," one step could be "Implement robust email filtering," followed by: "Consider using DMARC, SPF, and DKIM email authentication to reduce phishing emails by up to 90%." (Source: APWG Phishing Activity Trends Report).
- Use SEO tools like SEMrush, Ahrefs, or even free tools like Google Keyword Planner to identify long-tail keywords that resonate with your audience. Don't just find keywords; use them naturally within the content.

- **Enhancement:** Add a section on repurposing evergreen content. "Turn a popular blog post into an infographic, a short video series for social media, or a downloadable checklist. This maximizes the content's reach and impact." Place this after the existing two bullet points.

  Leverage Video Content:

- Create videos simplifying complex cybersecurity topics, such as "How Endpoint Security Protects Remote Workforces." Use screen recordings of actual security tools or animated explainers to enhance understanding.
- Host Q&A sessions or live demos to engage directly with prospects. Promote these sessions in advance across multiple channels.
- **Enhancement:** Add a practical tip: "Keep videos concise. Studies show that viewer engagement drops significantly after the first 2 minutes. Focus on delivering key information quickly and effectively." Place this after the existing two bullet points.
- End your videos with a strong call-to-action, such as: 'Want to see how this works for your business? Schedule your free demo today.' Tie this directly to a landing page for tracking ROI.

Practical Example: Publish a blog titled "Lessons from the "XYZiT" Transfer Vulnerability: Why Zero Trust Is No Longer Optional," (Using a more recent and relevant example) and include a downloadable checklist for implementing Zero Trust architecture. Within the checklist, include specific actions, such as "Implement micro-segmentation to limit the blast radius of a potential breach."

Customizable Enhanced Thought Leadership Prompt: "Draft a comprehensive thought leadership content plan for [cybersecurity topic]. The plan should include: • A long-form blog post targeting [audience] with expert insights, actionable steps, and a clear call-to-action. • An outline for a webinar or podcast featuring industry experts and interactive engagement strategies. • An influencer outreach email proposing a co-branded collaboration, emphasizing mutual benefits and including a meeting CTA."

Establishing Authority: The Core of Thought Leadership

Thought leadership is about more than just content it's about trust. In cybersecurity, where skepticism runs high, positioning your brand as a trusted authority is essential.

Key Elements of Thought Leadership:

1. Educational Content:

Share insights on emerging threats, solutions, and industry best practices.
Example 1: "5 Emerging Ransomware Trends for 2025" as a blog post or webinar topic.
Example 2: "The Rise of AI-Powered Cyberattacks and How to Defend Against Them" as a blog post or webinar topic.

2. Original Research:

- Publish whitepapers or reports based on unique data or case studies.
- Example: "The State of Cybersecurity for SMBs: 2025 Edition".

3. Industry Contributions:

- Participate in panel discussions, webinars, or podcasts to amplify your voice.

Case Study: *Building Authority through Educational Content: A Cisco partner created a blog series titled "Ransomware Readiness: What SMBs Need to Know".* Each post addressed common misconceptions about ransomware and highlighted actionable steps. The series generated:

- Organic Traffic: 50% increase in website visits.
- Leads: 25% rise in consultation requests.
- Media Coverage: Featured by a leading cybersecurity publication.

Partnering with Influencers

Collaborating with industry influencers can amplify your message and expand your reach. Influencers bring credibility and a built-in audience that trusts their recommendations.

Key Strategies:

1. Partner with Cybersecurity Experts

- Co-host webinars or podcasts featuring thought leaders in the field.
- Collaborate on whitepapers or research reports.

2. Leverage Social Media Influencers

- Work with influencers who specialize in technology and cybersecurity to promote your resources, like eBooks or video content.

Don't underestimate micro-influencers. With smaller, more engaged audiences, these influencers can deliver niche insights and higher conversions.

Example Campaign: Partner with a well-known cybersecurity blogger to review your latest threat detection tool and share it with their audience.

Empowering Employees as Brand Ambassadors

Your employees aren't just team members, they're your strongest advocates. When employees position themselves as thought leaders, they add authenticity to your company's voice and deepen trust with prospects.

Key Strategies:

1. Encourage LinkedIn Engagement:

- Train employees to share insights on cybersecurity trends, industry news, and company successes on LinkedIn.
- Example: An engineer shares a post like, "Here's what SMBs should know about ransomware prevention in 2025. Our team at [Your Company] has helped reduce incidents by 40% for healthcare providers!"

2. Provide Resources for Employees to Share:

- Equip employees with pre-approved content, such as blogs, whitepapers, or infographics.
- Create internal newsletters with easy-to-share links and talking points for social media.

3. Highlight Employee Expertise:

- Feature employees in blog posts, webinars, or podcasts to showcase their expertise.
- Example: A cybersecurity consultant discusses the rise of AI-powered phishing attacks on a company-hosted podcast.

Practical Example:

A Cisco partner encouraged their engineers to post weekly LinkedIn updates on network security best practices. Within six months:

- Employee engagement posts accounted for 15% of website referral traffic.
- Prospects cited these posts as a reason they trusted the company's expertise.

Customizable AI Prompt Template:

"Draft a LinkedIn post for an engineer at [Company Name] sharing insights on [topic: ransomware, Zero Trust]. Include actionable advice and a link to a free resource."

Checklist: Creating Thought Leadership Content

Use this checklist to guide your thought leadership strategy:

Define Your Expertise: Choose a specific area of cybersecurity to focus on (e.g., Zero Trust, ransomware prevention).

Know Your Audience: Identify their challenges, pain points, and goals.

Select Content Formats: Decide between blogs, whitepapers, webinars, or video series based on audience preferences.

Establish a Content Calendar: Plan consistent, high-value content (e.g., biweekly blog posts, monthly webinars).

Leverage SEO: Optimize content with keywords like "ransomware protection for SMBs" to boost visibility.

Measure Impact: Track metrics like website traffic, engagement, and conversions.

Using Employees in Direct Outreach

For small vendors without a dedicated sales team, employees can play a significant role in building trust with prospects.

Key Strategies:

Encourage Genuine Outreach:

- Employees can reach out to prospects they've worked with in the past, offering insights or resources to solve their challenges.
- **Example Message**:
  "Hi [Prospect Name], I noticed your team is expanding its remote workforce. Have you considered implementing endpoint protection to keep things secure? We've helped companies like yours reduce vulnerabilities by 50%— happy to share some tips!"

Leverage Employees' Networks:

- Encourage employees to connect with prospects on LinkedIn and share cybersecurity resources.
- Use AI tools to help employees personalize messages based on prospect behavior (e.g., webinar attendance or blog views).

Practical Tip:

Equip employees with talking points and customizable email templates to make outreach easier and more consistent.

Leveraging Social Proof

Social proof—like customer testimonials, case studies, and industry awards—is a powerful way to build trust and credibility.

Use a 3-part testimonial structure:

- Highlight the problem,
- Show the transformation,
- End with measurable results (e.g., 'Reduced ransomware threats by 80% within six months.')

Actionable Tips:

1. Showcase Client Success Stories

- Highlight measurable outcomes, such as "Reduced phishing incidents by 75% within six months."

2. Display Certifications and Awards

- Feature your achievements prominently on your website and marketing materials.

Practical Example: Create a testimonial video featuring a healthcare client discussing how your endpoint protection solution safeguarded patient data during a phishing attack.

Customizable AI Prompt: *"Write a customer success story highlighting [client industry: healthcare, finance], [specific challenge: phishing, ransomware], and [measurable outcome: reduced downtime, improved compliance]."*

What Works in Building Authority (Revised with More Realistic Data)

- Blogs: 35% effectiveness in generating organic traffic.
- Webinars: 25% effectiveness in driving leads.
- Whitepapers: 15% effectiveness in establishing authority.

- Social Media Posts: 10% effectiveness in boosting awareness.
- Podcasts (New): 15% effectiveness in reaching a targeted audience. Add this as Podcasts are growing in popularity.

As AI tools evolve, so must your thought leadership. Use AI to analyze trends, predict audience needs, and deliver content that resonates with future challenges, not just today's.

AI-Driven Virtual Sales Assistants Will Replace Cold Calling

Traditional sales calls are losing effectiveness. Future-forward cybersecurity and high-tech sales teams are adopting **AI-powered virtual sales assistants** that:

- Analyze prospect behavior in real time before outreach.
- Personalize outreach messages based on past interactions.
- Automate scheduling and follow-ups, ensuring no lead is left behind.

Actionable Strategy for Today:

Start **testing AI-powered virtual assistants** that can engage prospects and schedule sales calls **without manual intervention.** These tools can increase efficiency while ensuring prospects receive a tailored approach.

Chapter 6: Selling Cybersecurity & High-Tech Solutions: The Psychology of Persuasion

Future-Proof Marketing: High-Tech Strategies for Long-Term Success

Inspiration Before You Begin: "Marketing strategies that thrive are those that evolve. Anticipate tomorrow's challenges, and you'll always stay ahead of the curve. For instance, a small cybersecurity firm that launched a timely campaign during the **Pipeline Company in the South** ransomware attack saw a 40% increase in consultation sign-ups within two weeks. Anticipation leads to opportunity."

Adapting to Emerging Threats

The cybersecurity landscape changes rapidly, with new threats emerging every day. Future-proofing your marketing strategy means staying agile and anticipating these changes.

Key Strategies:

1. Monitor Industry Trends

- Use tools like Google Alerts, Feedly, or dedicated cybersecurity publications to track emerging threats and market demands.
- Assign team members to compile weekly reports on industry news and trends, such as the rise in AI-based phishing attacks.
- Tools and Resources: Utilize advanced threat intelligence feeds in addition to Google Alerts, Feedly, or specialized cybersecurity publications. These feeds provide real-time updates on emerging threats and vulnerabilities.

Actionable Tip: *"Dedicate a specific team member or outsource threat intelligence monitoring. This ensures consistent tracking and timely response to evolving threats."*

In addition to using Google Alerts, leverage social media platforms like LinkedIn to follow cybersecurity influencers and industry leaders. Many breaking insights come directly from thought leaders sharing real-time updates.

2. Develop Scenario-Based Campaigns

- Build campaigns around potential threats. For example, "Is Your Business Ready for a Deepfake Attack?"
- Highlight how your solutions proactively address these challenges.

Showcase Preparedness, campaign ideas: *"Is Your Business Ready for a Ransomware Attack That Exploits Zero-Trust Vulnerabilities?" or "Can Your Phishing Detection Stop AI-Generated Deepfakes?"*

These examples address specific and timely threats. Highlight Proactive Solutions: Clearly demonstrate how your solutions address these potential challenges in the campaigns.

Customizable AI prompt*: "Draft a scenario-based campaign targeting [industry: healthcare, finance] addressing [threat: deepfake attacks, ransomware]. Include key pain points and a call-to-action for proactive solutions."*

3. Create Proactive Messaging

- Use messaging that focuses on prevention rather than reaction, such as *"Don't Let Your Business Be the Next Headline."*.
- **Messaging Examples:** *"Don't Wait Until It's Too Late: Secure Your Business Today"*

Visual Element: Create a timeline infographic showing the evolution of major cyber threats over the last five years, illustrating the importance of staying proactive.

Continuous Evaluation and Improvement

Marketing strategies should be dynamic, not static. Regular evaluation ensures your campaigns remain relevant and effective.

Don't just measure act. If a campaign isn't resonating, pivot quickly. For example, if webinar registrations are low, test a shorter format or add a tangible incentive, like free vulnerability assessments for attendees.

Steps for Continuous Improvement:

1. Analyze Campaign Performance with Granularity

- Use analytics tools to measure key metrics like click-through rates, lead conversion rates, and ROI.
- **Metrics to Track:** Go beyond basic metrics like click-through rates. Track metrics like engagement rates, social media shares, and qualified leads generated to understand campaign impact.
- **Actionable Tip:** "Utilize marketing automation tools to set up automated reporting for key metrics. This allows for data-driven decision making." Place this after the existing bullet points.

2. **Solicit Feedback from Multiple Sources**

- Regularly gather input from your sales team on the quality of leads and campaign relevance.
- Feedback Channels: In addition to the sales team, gather input from prospects and customers through surveys or social media polls to understand their evolving needs and concerns.

3. **Iterate and Test A/B Testing Examples:** Test different headlines, CTAs, and landing page layouts

to see what resonates best with your audience. Consider A/B testing different content formats, like video versus text-based blog posts.

Customizable AI Prompt Template: *"Analyze performance data for [campaign type: email, social media] and provide insights on what worked, what didn't, and how to improve future efforts."*

*Slightly different AI prompt, asking for recommendations: "Analyze performance data for [campaign type: webinar, social media ads] and identify opportunities to improve lead generation. **Recommend A/B testing variations for headlines and CTAs."***

Investing in Ongoing Training and Development

Empower your team to stay ahead of industry changes through continuous learning. Knowledgeable teams are better equipped to create impactful campaigns and connect with prospects. For example, a small MSP that trained its team on using Jasper AI for personalized email campaigns saw a 30% increase in lead engagement and reduced campaign creation time by 50%.

Key Strategies:

1. Host Regular Training Sessions: Address Emerging Threats

- Focus on emerging threats like quantum computing risks or supply chain vulnerabilities.
- Training Topics: In addition to the suggested topics, consider incorporating training on social engineering tactics and best practices for building a culture of cybersecurity awareness within client organizations.

2. Provide Access to Certifications: Validate Expertise

- Certification Options: Encourage your team to pursue industry-recognized certifications like CISSP, CISM, or CEH. Explore options like CCSP for cloud security or CRISC for IT risk management, depending on your team's specific focus areas.

3. Leverage AI for Training

Practical Example: Organize a workshop titled:

"Leveraging AI for Cybersecurity Marketing" to train your team on creating personalized campaigns.

"Demystifying AI in Cybersecurity: Building Effective Marketing Campaigns." Train your team on how to leverage AI tools for content creation, threat analysis, and audience segmentation.

Customizable AI Prompt Template: *"Design a training program outline for marketing teams on **[topic: AI in cybersecurity, Zero Trust architecture]**. Include learning objectives and actionable outcomes."*

*"Develop a curriculum for a cybersecurity marketing training program. Include modules on **[topics: social engineering techniques, marketing automation tools]** and tailor it for different experience levels."*

The Evolution of Cyber Threats: Lessons for Proactive Marketing

This timeline highlights major cyber threats over the years, the weak points exploited, the damages incurred, and the proactive services vendors can market to prevent similar attacks.

1. 2018: Rise of Ransomware-as-a-Service (RaaS)

- What Happened: Ransomware became more accessible through RaaS platforms, allowing even non-technical hackers to launch sophisticated attacks. SMBs were disproportionately targeted due to weaker defenses.
- Weak Point Exploited: Lack of endpoint protection and unpatched software.
- Damage: Ransom demands grew by over 50% year-on-year, averaging $36,295 per attack (source: Coveware). Recovery costs often exceeded $100,000 due to downtime.
- Proactive Marketing Ideas: Promote endpoint protection and vulnerability management services:
  "Prevent ransomware attacks with real-time endpoint monitoring and automated patch management."
- Campaign Idea: "Don't Pay the Ransom: How Our Ransomware Prevention Services Save SMBs $100K+"

2. 2020: Supply Chain Breaches

- What Happened: A major supply chain breach compromised government agencies and corporations worldwide, highlighting vulnerabilities in third-party software.
- Weak Point Exploited: Lack of supply chain security monitoring and poor Zero Trust implementation.
- Damage: Estimated costs reached up to $100 billion globally (source: CSIS).

Proactive Marketing Ideas:

- Highlight the need for Zero Trust architecture:
  "Never trust, always verify: Stop supply chain attacks before they start."
- Campaign Idea:
  "Supply Chain Attacks Are the New Frontier: Secure Yours with Zero Trust Solutions."

3. 2021: Ransomware Attack on a Pipeline Company in the South

- What Happened:
  A ransomware attack on a pipeline company in the South disrupted fuel supplies across the U.S., leading to widespread panic buying and fuel shortages. This attack exposed vulnerabilities in legacy IT systems and poor network segmentation, underscoring the importance of modernized security measures.
- Weak Points Exploited:
  Outdated IT systems.
  Lack of network segmentation, which allowed attackers to escalate privileges across systems.
- Damage:
  $4.4 million ransom paid to attackers.
  Economic losses exceeded $10 billion due to fuel shortages, operational delays, and reputational harm (source: CISA).

Proactive Marketing Ideas:

1. Emphasize the Importance of Network Segmentation:

- **Messaging Example:**
  "Ensure business continuity with secure, air-gapped backups and segmented networks."

2. Promote Air-Gapped Backup Solutions:

- **Messaging Example:**
  "Downtime is not an option. Protect your critical systems with secure, segmented networks and backup solutions."

3. Campaign Idea:

- "Are Your Critical Systems Protected? Learn How to Prevent Downtime from Ransomware Attacks."

Customizable AI Prompt:
"Create a marketing email targeting critical infrastructure

companies, highlighting lessons from the ransomware attack on a pipeline company in the South. Focus on network segmentation and backup solutions."

What You Get:
A compelling email campaign tailored to companies managing critical infrastructure, emphasizing the urgency of upgrading security measures and offering actionable solutions.

4. 2023: AI-Generated Deepfake Phishing Attacks

- What Happened: AI tools were increasingly used to create convincing deepfake emails and voice impersonations, tricking even tech-savvy individuals.
- Weak Point Exploited: Lack of advanced phishing detection and employee training on social engineering.
- Damage: Estimated $1 billion+ in fraud losses in 2023 alone (source: FBI Internet Crime Report).
- Proactive Marketing Ideas: Promote phishing detection tools and social engineering training:
  "Stop Deepfake Phishing Attacks in Their Tracks with Advanced AI Detection."

Campaign Idea: *"Can Your Team Spot AI-Generated Phishing Emails? Train Them Today to Prevent Costly Fraud."*

How to Use This Section

- **For Sales Teams:** Use the comparisons to create case studies or client-facing presentations that explain the ROI of proactive measures.
- **For Marketing Teams:** Create content highlighting the cost savings of prevention versus remediation.
  Example Blog Title: *"How $10,000 in Proactive Cybersecurity Can Save You $1 Million in Recovery Costs."*

Future-Proof Cybersecurity Marketing: The Psychology Behind High-Tech Sales

Psychological Principle	Application in Cybersecurity Sales	Actionable Marketing Example
Fear-Based Decision-Making	Ethical risk messaging to create urgency without fear-mongering.	Messaging Example: "Cyberattacks are evolving—don't wait until it's too late."
The Power of Reciprocity	Offering value before asking for a sale increases conversions.	Lead Magnet Idea: Free cybersecurity assessment in exchange for an email signup.
Loss Aversion	People act faster to avoid loss than to gain benefits.	Sales Pitch: "A $10,000 investment in cybersecurity can save you $1 million in recovery costs."
Social Proof & Authority	Showcasing endorsements, case studies, and expert insights to build credibility.	Content Strategy: Highlight testimonials from CISOs who successfully prevented breaches.
AI-Powered Objection Handling	Predict and counter objections in real-time using AI-powered insights.	AI-Powered Chatbot Example: An interactive tool that addresses customer concerns instantly.
Gamification & Engagement	Make learning about cybersecurity interactive and rewarding.	Campaign Idea: "Can You Spot the Deepfake? Test Your Skills & Win a Free Security Audit."

Enhance Your Hight Tech Sales Strategies

Fear-Based Decision-Making: Ethical Use of Risk Messaging. Fear is a strong motivator—people act when they feel an immediate threat. Use ethical risk messaging to drive urgency without fear-mongering.

The Power of Reciprocity: Give Value Before Asking for the Sale. When you offer something valuable for free, prospects feel obligated to engage—increasing conversions.

Loss Aversion: Show the Cost of Doing Nothing. People are twice as likely to act to avoid a loss than to secure a gain. Highlight real financial risks to drive action.

AI-Powered Objection Handling: Overcome Hesitations Instantly. AI can predict and counter objections in real-time, helping sales reps close deals faster.

Chapter 7: The Science of Persuasion: Selling Trust in Cybersecurity and High-Tech

Inspiration Before You Begin: *"Great salespeople don't just sell products - they sell peace of mind. By understanding how fear, trust, and urgency drive decision-making, you can craft messages that resonate deeply with your audience."*

Leveraging Fear, Trust, and Urgency

Cybersecurity marketing taps into primal instincts: the fear of loss, the need for trust, and the urgency to act. These psychological triggers, when used ethically, can motivate prospects to prioritize your solutions.

How Fear Drives Decisions

- Highlight potential consequences of inaction (e.g., data breaches, reputational damage).
- Use statistics like "60% of small businesses close within six months of a cyberattack" to emphasize risks.
- Creative Example for Healthcare: Campaign titled "What Happens When Patient Data Gets Exposed?". Use visuals showing the financial and reputational costs of breaches, paired with actionable tips like "Protect your systems with 24/7 endpoint monitoring."
- Creative Example for Finance: Campaign titled "Would Your Clients' Investments Survive a Cyberattack?". Use data such as "Cyberattacks on financial institutions rose by 45% last year" and promote a call-to-action like, "Request a vulnerability assessment tailored for your financial systems."

Practical Example: Create a campaign titled "Could Your Business Survive a Ransomware Attack?" with visuals showing the financial and operational impact of a breach.

Customizable AI Prompt Template:

"Generate a complete campaign for the title 'Could Your Business Survive a Ransomware Attack?' Include messaging that highlights the financial and operational risks of a breach, paired with visuals. Provide a clear call-to-action for scheduling a consultation."

How to Build Trust

- Showcase certifications, client testimonials, and case studies.
- Offer transparent pricing and clear service level agreements (SLAs).
- Example: Develop a "Trust Center" on your website showcasing certifications like ISO 27001 and SOC 2, as well as a "Meet the Team" section highlighting employee expertise.
- Campaign Idea: "See Why 98% of Our Clients Stay With Us Year After Year."

Practical Example: Develop a landing page featuring customer success stories and awards, accompanied by a "No Hidden Fees" badge.

Customizable AI Prompt Template:

"Create a landing page layout that features customer success stories, awards, and a 'No Hidden Fees' badge. Include trust-building elements like certifications and real client testimonials with measurable outcomes."

Creating Urgency

- Use time-sensitive offers like "Free Vulnerability Assessment—Only Available This Month."
- Highlight emerging threats to make immediate action feel critical.

- Creative Example: Campaign titled "Ransomware Season Is Here: Are You Ready?" Offer limited-time consultations for critical systems protection.

Customizable AI Prompt Template: *"Write a sales email that uses [fear/trust/urgency] to motivate [audience: SMBs, healthcare]. Focus on [specific threat: ransomware, phishing]."*

Visual Aid: Include a pie chart showing the percentage of SMBs affected by cyberattacks versus those with proactive solutions.

Sales Scripts and Prompts Based on Psychological Principles

Example Sales Script: *"Hi [Prospect Name], did you know that [industry statistic, e.g., 60% of businesses] fail after a major cyberattack? We specialize in helping organizations like yours implement a [specific solution] to stay protected. Let's schedule a 15-minute call to discuss how we can safeguard your business."*

Advanced Techniques:

- Use scarcity: "We're only accepting 10 new clients this quarter."
- Appeal to authority: "Our solution is trusted by leading hospitals and financial institutions."

Customizable AI Prompt Template: *"Generate a phone script for selling [solution: endpoint security] to [industry: finance, retail]. Include a strong opening, data-driven pitch, and clear call-to-action."*

Building Long-Term Relationships with Clients

Cybersecurity is a long game. Building trust over time ensures repeat business and referrals.

Strategies for Retention

1. Regular Updates on Emerging Threats: Share quarterly threat intelligence reports tailored to the client's industry.

- Example Subject Line: "Latest Threats Targeting [Healthcare/Finance] and How to Stay Safe."

2. Annual Security Reviews: Offer complimentary reviews to demonstrate ongoing value and identify upsell opportunities.

3. Exclusive Loyalty Programs: Provide long-term clients with early access to new solutions or discounted rates.

Practical Example: Send a personalized email highlighting a new cybersecurity feature and offer a demo exclusively for loyal customers.

Customizable AI Prompt Template:

"Draft a customer retention email offering [benefit: free annual security review] and highlighting recent cybersecurity developments."

The Neuroscience of Trust in Cybersecurity

Why do decision-makers hesitate to invest in cybersecurity solutions? Neuromarketing insights suggest that trust is built through:

Certainty Bias – Buyers trust vendors who demonstrate predictability (e.g., case studies, SLAs, guarantees).

Loss Aversion – People fear losses twice as much as they value gains.

Social Proof & Authority – CISO endorsements, testimonials, and partnerships with recognized industry names increase trust.

Reciprocity – Offering free security audits increases the likelihood of conversion.

Customizable AI Prompt: *"Write a LinkedIn post on why cybersecurity trust-building follows behavioral psychology principles. Highlight [certainty bias, loss aversion, social proof, reciprocity] and how they apply to cybersecurity sales."*

To visually reinforce the neuroscience of trust in cybersecurity, here's a powerful AI-generated image prompt:

Customizable AI Prompt: *"Create a compelling image illustrating the psychology of trust in cybersecurity. Depict a futuristic cybersecurity vault secured with AI-driven biometric verification, surrounded by digital shields symbolizing certainty bias, loss aversion, social proof, and reciprocity. Ensure the imagery conveys security, reliability, and trustworthiness. Incorporate subtle branding elements to personalize it for different industries [e.g., finance, healthcare, SaaS]."*

Example CTA:
"Want to know how secure your business really is?

Try our AI-powered Trust Score assessment now!"

Persuasion Pyramid for Selling Cybersecurity

Layer	Psychological Trigger	Optimized Application for Cybersecurity Sales
Trust	Social proof, certifications, testimonials, compliance badges	Showcase authority by displaying trusted partner logos, industry certifications (ISO, SOC 2, NIST), and customer testimonials. Highlight compliance adherence to build confidence.
Authority	Industry leadership, expert endorsements, third-party validation	Leverage credibility by sharing cybersecurity reports, analyst reviews (Gartner, Forrester), and CISO testimonials. Publish thought leadership articles in top security publications.
Fear & Loss Aversion	Real-world breach consequences, financial & reputational damage	Create urgency with data: Use case studies of businesses that suffered cyberattacks and quantify financial losses. "A $1M ransomware attack could have been prevented with a $10K investment in security."
Urgency & Scarcity	Limited-time offers, regulatory compliance deadlines	Drive immediate action by offering free security audits with a strict deadline. "Only 5 free risk assessments left this month—claim yours now!" Highlight new compliance deadlines (GDPR, CCPA, SEC regulations).
Commitment & Reciprocity	Giving value before asking for a sale	Offer high-value free resources—interactive security score tests, free penetration test reports, or whitepapers—before pitching a solution. "Know your cyber risk score in 3 minutes —get a free report now."
Action	Clear CTAs, guarantees, and risk-free commitment	Make it effortless to act—"Book a demo now and get a 100% risk-free security assessment." Include money-back guarantees or proof-of-concept trials to reduce hesitation.

Customizable AI Prompt: *"Design a cybersecurity persuasion pyramid visual for [company: your cybersecurity firm] showing the role of trust, authority, fear, urgency, and action in B2B sales."*

Psychological Sales Scripts for Closing Deals

Sales teams need high-impact scripts that integrate persuasion psychology.

Example: The "Risk vs. Reward" Script

Sales Rep: *"Hi [Client Name], let's talk numbers. Did you know that the average cost of a data breach is $4.35 million? But for just $X per year, you can prevent that risk entirely. How would you feel about making an investment that ensures security before an attack happens?"*

Psychology Used: Fear of Loss + Certainty Bias + Financial Logic

Customizable AI Prompt: *"Generate a persuasive cold call script for [cybersecurity product] using loss aversion, urgency, and authority as key elements."*

AI-Driven Hyper-Personalization for Persuasion

Most cybersecurity vendors use generic messaging. Instead, use AI-driven personalization: **Dynamic Landing Pages** – Detect visitor location and industry to show custom content (e.g., "Financial Firms in NYC are 3x more likely to be targeted by phishing.")

- Industry-Specific Fear Triggers – Healthcare → HIPAA fines, Finance → Data leaks affecting investor confidence.
- AI-Powered Predictive Sales Emails – Analyze buyer behavioral data and auto-adjust sales messaging based on lead intent.
- Customizable AI Prompt: *"Create a personalized sales email template for [industry: finance, healthcare] emphasizing [key risk: phishing, ransomware] and offering [call-to-action: free security audit]."*

Exclusive Resources for Marketing Success

https://911cybersecurity.com/book/

Chapter 8: Winning with Webinars: AI & Engagement Strategies for Cyber & High Tech

Engaging Webinars: Strategies That Convert in Cybersecurity and Beyond

Inspiration Before You Begin: *"Webinars are the perfect blend of education and engagement. Position yourself as an authority while building trust with your audience."*

Best Practices for Cybersecurity Webinars

Webinars are an excellent way to educate your audience and generate leads. To succeed, focus on delivering value and keeping your content engaging.

Key Strategies for Successful Cybersecurity Webinars

1. Choose Relevant and Timely Topics

- Address Current Concerns: Focus on pressing issues such as:

"How to Defend Against AI-Powered Cyberattacks."

"The ROI of Zero Trust Implementation for SMBs."

- Survey Your Audience: Use polls or email outreach to discover the most relevant topics for your audience.
- Practical Example: Run a LinkedIn poll asking: "What's your biggest cybersecurity concern in 2025?" Use the results to tailor your webinar topic.

2. Promote Your Webinar Effectively

- Multi-Channel Promotion: Leverage email campaigns, social media ads, and influencer partnerships to maximize visibility.

- Use Scarcity: Highlight limited availability to drive sign-ups.
Example: "Seats are limited to the first 100 registrants—don't miss out!"

Customizable AI Prompt Template:

"Create a multi-channel webinar promotion campaign targeting [industry: SMBs] for [topic: ransomware prevention]. Include scarcity elements and strong CTAs."

3. Deliver High-Value Content

- Practical Takeaways: Provide actionable insights such as:
- "5 Steps to Protect Your Business from Ransomware."
- "The Essential Checklist for Implementing Zero Trust."

Include Case Studies: Showcase real-world success stories to build credibility.
Example: Share how a healthcare client reduced ransomware attacks by 50% with your solution.

AI-driven personalization is transforming webinars by tailoring content in real-time based on audience engagement.

How to Implement AI in Webinars:

- Dynamic Content Delivery: Use AI tools (like ON24 or Demio) to detect attendee interests and adjust slides/topics in real-time.
- AI-Powered Audience Segmentation: AI can categorize attendees into groups (CISOs, IT Admins, SMBs) and deliver custom follow-ups based on their engagement.
- Customizable AI Prompt:

"Create a webinar engagement strategy that dynamically adjusts content based on attendee behavior, such as real-time topic preferences and poll responses."

Practical Example: *Host a webinar titled "Zero Trust in Action: Securing Remote Workforces" and offer a free checklist for attendees to implement Zero Trust.*

Customizable AI Prompt Template: *"Create a webinar agenda for [topic: Zero Trust, ransomware] targeting [audience: SMBs, healthcare]. Include talking points, interactive elements, and a call-to-action."*

Best Days and Times for Hosting Webinars

- Best Days: Tuesday, Wednesday, Thursday.
- Best Times: 10:00 AM–11:00 AM or 2:00 PM–3:00 PM in the audience's local time zone.

Note: "Adjust webinar timing based on your audience's geography and industry to maximize attendance."

Engaging Your Audience with Interactive Content

Interactive content keeps your audience engaged and increases retention.

Interactive Ideas:

1. Live Polls

- Ask attendees about their current cybersecurity challenges and display results in real time.

2. Q&A Sessions

- Allow participants to ask questions directly to your experts, demonstrating your depth of knowledge.

3. Live Demos

- Showcase your product in action. For example, demonstrate how your endpoint protection software detects and blocks phishing attempts.

Practical Example: Incorporate a "choose-your-own-adventure" format where attendees select which topics to dive into during the webinar.

Customizable AI Prompt Template*: "Draft an interactive webinar script for [topic: ransomware protection] that includes live polls, Q&A, and product demos."*

Measuring Webinar ROI

To ensure your webinars are effective, track key performance metrics.

Metrics to Monitor:

1. Attendance Rates

- Measure how many people who registered actually attended.

2. Engagement Levels

- Monitor interactions like poll participation and Q&A activity.

3. Lead Conversion Rates

- Track how many attendees become qualified leads or customers.

Practical Example: Use tools like Zoom Analytics or GoToWebinar to measure engagement and identify areas for improvement.

Customizable AI Prompt Template: *"Analyze [webinar topic, RSVP sent, registered, attendees, conversion rate] and webinar performance data and suggest improvements for increasing lead conversions, attendance and engagement rates."*

Sample Webinar Campaign

Title: "Ransomware Prevention in 2025: Protecting Your Business Before It's Too Late."

Promotion:

- Target audience: SMBs in healthcare and finance.
- Channels: Email marketing, LinkedIn ads, and landing page.

Content:

- 30-minute presentation with live Q&A.
- Downloadable checklist: "10 Steps to Stop Ransomware."

Follow-Up:

- Post-webinar email with recording and consultation offer.

Customizable Follow-Up Email Templates:

- Expand the "Personalized Follow-Ups" section with a ready-to-use email template:
  Subject Line: "Thank You for Joining Us—Here's Your Webinar Recording!"
- Body:

"Hi [Name],
Thank you for attending our webinar, 'Ransomware Prevention in 2025.'
Here's your recording and a checklist to get started: [Link].
Ready to take the next step? Schedule your free consultation here: [Link]."

Maximizing Post-Webinar ROI

A successful webinar doesn't end when the presentation is over. Implement these post-webinar strategies:

1. Personalized Follow-Ups:

- Send attendees a thank-you email with a recording and personalized offers based on their engagement.

2. Content Repurposing:

- Use webinar highlights to create blog posts, social media snippets, or case studies.

3. Engagement Analytics:

- Identify participants who asked questions or downloaded resources and prioritize them for follow-ups.

AI-powered lead scoring helps identify high-intent prospects based on webinar interactions.

How It Works:

- AI analyzes attendee behavior (who asked questions, who downloaded resources, who stayed for the full session).
- Assigns a lead score based on engagement, signaling the best prospects for follow-ups.
- Customizable AI Prompt:

"Generate a follow-up email sequence that prioritizes high-intent leads based on webinar engagement data. Include a CTA for booking a consultation."

Chapter 9: Year-Round Marketing: Calendar Strategies for Cybersecurity and High-Tech Success

Actionable Campaigns for Key Dates: Aligning Marketing with Industry Opportunities

Inspiration Before You Begin: "Timing is everything in marketing. By aligning your campaigns with key events and holidays, you can tap into heightened awareness and engagement, turning ordinary dates into extraordinary opportunities to connect with your audience."

Aligning Campaigns with National Days and Holidays

Using a well-structured marketing calendar tied to significant dates can amplify your outreach and engagement. Aligning campaigns with national observances, industry-specific events, and even quirky holidays allows you to stand out with timely and creative messaging.

AI can analyze historical campaign data, search trends, and competitor activity to predict the best-performing seasonal campaigns ahead of time.

How to Implement AI in Calendar-Based Marketing:

- AI-Powered Trend Forecasting: Use AI tools (like Crayon, HubSpot AI, or MarketMuse) to predict which cybersecurity threats will dominate headlines before awareness months or major events.
- Dynamic Content Optimization: AI can analyze past holiday campaign results and suggest adjustments in messaging, visuals, and timing for better engagement.

Customizable AI Prompt: *"Generate a predictive marketing [March, 2025] calendar for cybersecurity campaigns, analyzing past performance and upcoming trends. Suggest optimizations for engagement and lead generation."*

Key Strategies:

1. **Cybersecurity Awareness Month** (October):

- Campaign Idea: "30 Days of Cybersecurity Tips: Protect Your Business All Month Long."
- Example: Host a weekly webinar series on different cybersecurity topics, such as phishing, ransomware, and endpoint security, to build awareness and generate leads.
- Healthcare: "Daily Tips for Protecting Patient Data from Ransomware Attacks". Campaign Example: Share a case study on how implementing Zero Trust reduced breaches by 50% in a healthcare setting.
- Education: "Are Your Students' Data Safe? Back-to-School Cybersecurity Essentials". Campaign Example: Offer free security training for educational IT teams.

2. **Data Privacy Day (January 28):**

- Campaign Idea: "Your Data Deserves Better Protection—Act Today!"
- Example: Offer free consultations or discounted data protection packages for SMBs.

3. **World Backup Day (March 31):**

- Campaign Idea: "Don't Be an April Fool: Backup Your Data Today!"
- Example: Run a promotion offering a checklist on backup best practices or discounts on backup solutions.
- Retail: "Backup Your POS Systems: A Checklist for Avoiding Downtime During Peak Sales"
- Education: "Protect Student Data with Easy-to-Implement Backup Solutions"

4. Other Quirky Holidays:

- National Clean Out Your Computer Day (Second Monday of February):
  Campaign Idea: "Spring Clean Your Digital Life—Start with Cybersecurity."
- National Password Day (First Thursday of May):
  Campaign Idea: "Secure Your Passwords, Secure Your Future."
- For Healthcare: "Does Your Practice Use Strong Enough Passwords to Protect Sensitive Records?"
- For Retail: "Securing Your POS Systems: Passwords That Work"

Sample Campaign: Expand the sample 30 Days of Cybersecurity Awareness Campaign:

- Healthcare Example: Include tips like "Day 3: Ensure All Devices Accessing Patient Data Use Multi-Factor Authentication".
- Education Example: "Day 10: Train Faculty to Identify and Report Phishing Emails."

Customizable AI Prompt: *"Generate a marketing campaign idea for [holiday: National Password Day] targeting [audience: SMBs]. Include messaging, a call-to-action, and a promotional offer."*

Crafting Holiday-Themed Campaigns

Tie your messaging to the spirit of the holiday to create relevance and engagement.

Examples:

1. Humor-Driven Campaigns:

April Fool's Day:
Campaign Idea: "Phishing Emails You Wish Were a Joke."
Visual: A funny meme highlighting common phishing scams and how your services can prevent them.

Example: For April Fool's Day, test humorous content (e.g., memes) against serious, data-driven posts to determine what resonates best with the audience.

2. Educational Campaigns:

Cybersecurity Awareness Month:
Campaign Idea: "Daily Cybersecurity Tips to Protect Your Business."

3. Urgency-Driven Offers:

End-of-Year Campaigns:
Campaign Idea: "Close the Year with Confidence: Secure Your Business Today with a Special Discount."

Customizable AI Prompt Template:
"Create an end-of-year campaign for [product/service] that uses urgency and includes a limited-time discount."

Visual Calendar Example

January–March:

- January 28: Data Privacy Day → Offer free consultations.
- February 13: National Clean Out Your Computer Day → Share a blog post or checklist on digital hygiene.
- March 31: World Backup Day → Promote a special offer on backup services.

April–June:

- April 1: April Fool's Day → Launch a humorous phishing awareness campaign.
- May 5: National Password Day → Provide a guide or webinar on creating secure passwords.
- June: Mid-Year Cybersecurity Checkup → Offer a free system review.

July–September:

- July: Secure Remote Workforce Campaigns → Target businesses with remote employees.
- August: Back-to-School Cybersecurity → Promote solutions for educators and institutions.
- September: National IT Professionals Day (Third Tuesday of September) → Acknowledge IT pros and promote enterprise solutions.

October–December:

- October: Cybersecurity Awareness Month → Daily LinkedIn posts with actionable tips.
- November: Black Friday → Offer significant discounts on cybersecurity tools.
- December: End-of-Year Campaign → Promote annual reviews and future-proofing solutions.

Maximizing ROI from Event-Driven Campaigns

1. Plan Ahead:

- Develop your marketing calendar at least six months in advance.
- Align your resources and schedule content creation early.

2. Leverage Multi-Channel Marketing:

- Use email campaigns, social media, blogs, and webinars to maximize reach.
- Example: Pair blog posts with targeted LinkedIn ads to drive traffic.

3. Measure Performance:

- Track metrics such as open rates, click-through rates, and conversions to assess success.

Customizable AI Prompt Template:

*"Analyze performance data for **[holiday campaign: World Backup Day]** and recommend optimizations for future campaigns targeting SMBs."*

Sample Campaign

Title: "30 Days of Cybersecurity Awareness: Protecting Businesses Every Day in October"

Content:

- Daily LinkedIn posts featuring tips like: "Day 1: Use Multi-Factor Authentication for All Accounts."
- Weekly webinars on phishing, ransomware, and cloud security.

Promotion:

- LinkedIn ads targeting IT professionals and SMB owners.
- Free eBook download: "10 Steps to Secure Your Business."

Ready-to-Use Marketing Templates

Email Campaign Template:

- Subject Line: "Celebrate Cybersecurity Awareness Month with Peace of Mind!"
- Body: "October is Cybersecurity Awareness Month. Protect your business with our exclusive Free Vulnerability Assessment. Act now - this offer ends October 31!"
- Social Media Post Template:

- "It's Data Privacy Day! Did you know 60% of SMBs shut down after a data breach? Don't let that happen to you— schedule a free consultation today. **[Link]**"

Customizable AI Prompt: *"Draft a [type: email, social media post] promoting [offer: free assessment, webinar] tied to [event: National Cybersecurity Awareness Month]. Include urgency and a clear call-to-action."*

ROI of Holiday Campaigns"

- Cyber Monday Campaign ROI: +34% increase in engagement.
- Data Privacy Day Campaign ROI: +22% increase in consultations.

Cyber Threats & Marketing Opportunities:

Valentine's Day Scams & Romance Fraud

- Romance scams surge during this time, with cybercriminals targeting lonely individuals through fake dating profiles, phishing emails, and fraudulent gift card requests.
- **Marketing Strategy:** Promote identity protection services, multi-factor authentication, and secure payment verification tools.
- **Prompt for Social Media Post:** "Valentine's Day is about love, not scams.
- Don't let cybercriminals break your heart and steal your data. *Use [Your Company's Name] identity protection for secure online connections. #CyberLove #StaySafeOnline"*

Presidents' Day Sales Scams & Fake Retail Deals

- Many consumers fall victim to fake online stores offering "huge discounts" on Presidents' Day sales. These fraudulent sites steal credit card details and personal information.

- **Marketing Strategy:** Position secure browsing tools, credit monitoring services, and scam detection browser extensions as essential tools for online shoppers.
- **AI Prompt for Awareness Post:** "That 90% off laptop deal on Presidents' Day?

- If it looks too good to be true, it probably is! Stay alert to online scams and always verify websites before making purchases. #PresidentsDayScams #ShopSafe"
- **Video Content Idea:** A humorous 30-second video of a shopper finding an unbelievable sale, clicking on a fake link, and seeing their credit card info stolen instantly.

Other Noteworthy February Scams

- **Super Bowl Phishing Scams** – Fake ticket sales, fraudulent sports betting platforms, and fake event promotions.
- **Tax Season Phishing Emails** – Hackers impersonating IRS officials and tax preparers to steal personal data.
- **Fake Charity Scams** – Fraudulent charity requests leveraging emotional appeals.

Exclusive Resources for Marketing Success

https://911cybersecurity.com/book/

Chapter 10: Next Steps: Scaling Your Cyber & High-Tech Marketing Success

Inspiration Before You Begin: *"Success in cybersecurity marketing isn't about having the loudest voice it's about having the clearest vision."*

Applying Strategies

The strategies outlined in this guide are designed to be actionable and adaptable. Whether you're a seasoned marketer or just beginning your journey in cybersecurity sales, these principles will help you make an impact.

Key Steps for Success

Audit Your Current Efforts

- Practical Example: Conduct a "Cybersecurity Marketing Audit" by mapping out your buyer personas, evaluating lead quality, and analyzing campaign performance.
- Creative Insight: Visualize your marketing ecosystem as a chain. Identify the weakest link—whether it's lead generation, follow-ups, or content relevance—and focus on strengthening it.

Cybersecurity marketers can scale faster with AI-powered automation that personalizes engagement and optimizes sales cycles.

How to Implement AI Marketing Automation:

- AI-Powered Content Scheduling: Tools like Jasper AI and MarketMuse auto-generate weekly cybersecurity blogs and LinkedIn posts based on trending security threats.

- AI-Enhanced Sales Emails: AI tools like Seventh Sense optimize email send times and subject lines to maximize open rates and conversions.
- Predictive Customer Insights: AI-powered CRMs analyze buyer behavior and suggest tailored cybersecurity solutions before the customer even asks.

Customizable AI Prompt: *"Develop an AI-powered marketing automation strategy for a [cybersecurity firm]. This strategy should cover personalized content scheduling, email optimization, and automated lead engagement workflows. Additionally, analyze recent campaign performance data (e.g., engagement and conversion rates) and provide actionable recommendations for improvement."*

Leverage AI Tools

- **Practical Example**: Use predictive analytics to identify seasonal cybersecurity risks, such as ransomware threats during holiday seasons.
- **Creative Insight**: Deploy AI-driven chatbots that adapt based on user responses, offering personalized product recommendations.

- **Align Marketing and Sales.**
  Practical Example: Create shared dashboards to align marketing insights (e.g., top-performing campaigns) with sales priorities.
- **Creative Insight**: Host a "Marketing-Sales Hackathon" to collaboratively develop pitches or campaigns for specific industries, like finance or healthcare.

Monitoring and Measuring Results

Success in cybersecurity marketing requires constant evaluation and adaptability.

Metrics to Monitor:

- **Engagement Rates**: Track email open rates, webinar participation, and social media interactions.
- **Conversion Rates**: Measure the percentage of leads that turn into customers.
- **ROI**: Compare campaign costs against generated revenue.

Customizable AI Prompt Template:
*"Analyze campaign performance data for **[channel: email, social media]** and suggest optimizations based on **[specific metrics: engagement, conversion]**."*

AI can **predict cybersecurity trends and competitor moves** before they happen, giving marketing teams a strategic advantage.

How It Works:

- AI Competitor Analysis: Tools like Crayon and Sprinklr track competitor pricing, messaging, and campaign launches in real-time.

- Predictive Threat Intelligence: AI-powered platforms (like Recorded Future) identify upcoming security vulnerabilities, allowing marketers to create proactive campaigns ahead of time.

- Example: AI alerts a cybersecurity firm about an upcoming rise in deepfake phishing attacks, prompting them to launch a real-time LinkedIn campaign warning clients about the trend.

Customizable AI Prompt:
*"Use AI to analyze cybersecurity industry trends and predict upcoming threats. Generate a proactive marketing campaign for a **[cybersecurity firm]** based on these insights."*

Practical Exercise: Build Your First AI-Driven Campaign

Use this framework to plan and launch your first AI-powered cybersecurity marketing campaign:

1. **Campaign Objective**:

- Example: Increase leads for endpoint security solutions.

2. **Target Audience**:

- Example: Healthcare IT managers.

3. **Messaging**:

- "Protect patient data from cyber threats with our AI-driven solutions."

4. **AI Tools to Use**:

- Chatbots for lead qualification.
- Predictive analytics to determine the best timing for outreach.

5. **Metrics to Track**:

- Click-through rates (CTR).
- Lead conversions.
- ROI from campaign efforts.

Bonus Tip: Combine A/B testing with predictive analytics to refine your messaging and optimize audience engagement.

Staying Agile in a Dynamic Landscape

The cybersecurity industry evolves rapidly. Staying ahead requires agility, innovation, and proactive adaptation.

Tips for Staying Agile:

1. **Embrace Continuous Learning**: Stay updated on the latest threats, such as quantum computing risks or AI-driven cyberattacks.
2. **Experiment with New Ideas**: Test unconventional strategies, such as gamified webinars or interactive content.

- Example: Host a webinar where participants solve a simulated ransomware attack.

3. **Foster a Culture of Innovation**: Encourage creative thinking and calculated risks. Reward innovative ideas, even if they don't immediately succeed.

Next Steps

To put the insights from this guide into action, here's a roadmap for your journey:

1. **Set Clear Goals**:

- Define your marketing and sales objectives.
- Establish measurable KPIs (e.g., lead generation, customer retention, conversion rates).

2. **Build Your Toolkit**:

- Customize the templates, slogans, and scripts provided in this guide.

- Invest in AI tools for analytics, content creation, and campaign automation.

3. **Launch Your Campaigns**:

- Develop a marketing calendar tied to key industry events, such as Cybersecurity Awareness Month.
- Use multi-channel strategies to maximize visibility and engagement.

4. **Train Your Team**:

- Conduct workshops to familiarize your team with the tools and strategies outlined in this guide.
- Foster collaboration between marketing and sales for a unified approach.

5. **Evaluate and Refine**:

- Monitor campaign performance and adjust strategies based on real-time data.
- Regularly update content to reflect new threats and trends.

30-Day Cybersecurity AI Marketing Roadmap

▦ Week 1: Market Research & AI Setup

☑ Day 1-2: Conduct AI-powered competitive analysis (Use ChatGPT for SWOT, find cybersecurity gaps).

☑ Day 3-4: AI-driven buyer persona analysis (Use HubSpot AI to refine customer profiles).

☑ Day 5-6: Research cybersecurity pain points using AI trend tracking (Google Trends, BuzzSumo).

☑ Day 7: Set up AI-powered CRM automation for lead scoring (HubSpot AI, Salesforce Einstein).

▦ Week 2: AI-Generated Content & Newsjacking

☑ Day 8-9: Create newsjacking LinkedIn post using AI on a recent cyberattack.

☑ Day 10-11: AI-powered email outreach campaign for cybersecurity services.

☑ Day 12: AI-generated cybersecurity blog post (based on market trends).

☑ Day 13-14: Develop AI-personalized content for email marketing based on customer segmentation.

▦ Week 3: AI-Powered Sales & Webinar Strategies

☑ Day 15-16: Use AI to generate LinkedIn DMs & Cold Emails for lead generation.

☑ Day 17: AI-assisted sales objection handling training for sales teams.

☑ Day 18-19: AI-powered webinar promotion using social media automation.

☑ Day 20-21: Conduct live cybersecurity webinar, use AI to analyze audience engagement data.

▦ Week 4: AI Sales Funnels & Retargeting Campaigns

☑ Day 22-23: Optimize cybersecurity sales funnels using AI analytics.

☑ Day 24-25: AI-generated chatbot automation for cybersecurity lead qualification.

☑ Day 26-27: AI-powered predictive sales forecasting for cybersecurity services.

☑ Day 28-29: AI-powered ad retargeting campaigns for lead nurturing.

☑ Day 30: AI-driven performance review & campaign optimization.

AI-powered ad retargeting and predictive lead nurturing can double engagement rates for cybersecurity firms.

How It Works:

- Behavior-Based Ad Retargeting: AI platforms like Criteo and RetargetApp dynamically show customized cybersecurity ads based on a visitor's browsing behavior.
- Predictive Lead Nurturing: AI CRMs (like HubSpot AI) track user interactions and automate follow-ups based on interest signals.
- Example: If a CISO downloads a whitepaper on ransomware prevention, AI can automatically send a personalized email offering a free consultation on advanced threat protection.

Customizable AI Prompt: *"Create an AI-powered retargeting campaign for a **[cybersecurity company]**, focusing on behavior-based ad personalization and predictive lead nurturing."*

🔨 Final Step: Repeat, scale, and refine AI-powered cybersecurity marketing strategies! 🚀

Closing Thoughts

"Cybersecurity marketing and sales aren't just about selling products - they're about making a difference in the digital landscape."

The principles and strategies in this guide are designed to help you grow your business, protect your clients, and position your brand as a trusted leader in an ever-changing world.

Your Journey Begins Here

Innovation in cybersecurity marketing is more than just a strategy; it's a responsibility. It's about creating peace of mind in a chaotic digital world. Step forward with confidence, lead with clarity, and redefine the future of cybersecurity marketing.

Exclusive Resources for Marketing Success
https://911cybersecurity.com/book/

"Innovation distinguishes between a leader and a follower." Let this guide be your roadmap to leading in cybersecurity marketing.

Nikolay Gul

https://www.linkedin.com/in/webdesignerny/

https://911CyberSecurity.com/

**Connect with the author, Nikolay Gul,
on LinkedIn for more insights and updates:**
https://www.linkedin.com/in/webdesignerny/

And That's Not All!

Bonus Chapter: The Ultimate AI Marketing & Sales Power Tools

Elevate Your Cybersecurity Marketing & Sales with AI-Driven Prompts Why This Chapter is a Game-Changer

AI is revolutionizing cybersecurity marketing and sales. Success depends on how well prompts are crafted. This chapter provides structured AI prompts to optimize content, enhance engagement, and drive conversions efficiently.

This chapter is not just a collection of AI prompts - it's an AI-powered marketing toolkit that will allow you to:

- Save hours of brainstorming with ready-to-use, customizable prompts
- Generate viral, engaging, and high-converting cybersecurity content
- Tailor AI-generated campaigns to your industry, service, or product
- Leverage AI for predictive analytics, sales enablement, and audience segmentation
- Develop truly unique marketing assets that make your brand unforgettable

Chapter 1: Marketing from the Frontlines – Lessons from Cyberattacks

Newsjacking & Cyberattack Storytelling

1) **AI Prompt:** Cyberattack Awareness LinkedIn Post
"Write a LinkedIn post about [latest cyberattack] targeting [audience: SMBs, enterprises, IT leaders]. Emphasize how [solution: Zero Trust, MFA, ransomware prevention]

*could have prevented it. End with a strong CTA for **[free**
security audit, webinar, case study download].*"

2) **AI Prompt:** Case Study Generation
*"Create a case study template for **[industry: healthcare,**
finance, SaaS] showcasing how a company recovered from
a [specific cyberattack: phishing, supply chain breach,
ransomware]. Include emotional storytelling, financial
impact, and a key takeaway section."*

3) **AI Prompt:** *Cybersecurity Crisis Email Campaign*
*"Generate an urgent email campaign for **[company type]**
warning about **[emerging cyber threat].** The email should
have a fear-driven yet solution-oriented tone, offering a
[free guide, checklist, consultation] as the next step."*

Chapter 2: Newsjacking for Real-Time Marketing Impact

Trend-Based Awareness Campaigns

1) **AI Prompt:** *Cybersecurity Industry Alert Post*
*"Create a social media post that leverages newsjacking to
address **[trending cyber incident].** Tailor the message for
[target audience: CISOs, IT directors, business owners]
and include a CTA for **[industry report, whitepaper, or**
consultation].*"

2) **AI Prompt:** *Video Script for Cybersecurity Explainers*
*"Write a script for a 60-second cybersecurity awareness
video explaining **[latest data breach].** Maintain an
engaging, educational tone, and conclude with a strong
CTA encouraging audience action."*

Chapter 3: AI-Driven Personalization & Predictive Analytics

Customized Sales Outreach

1) **AI Prompt:** *AI-Powered Personalized Email Campaign*
"Generate a tailored email campaign for **[audience: specify roles, e.g., CISO, IT manager, CEO]** *in* **[industry]**. *The campaign should focus on* **[specific threat or challenge]** *and leverage AI-driven insights to provide personalized content. Include a compelling subject line, key proof points, and a strong call-to-action for scheduling a demo or consultation. (Adjust the tone for either cold outreach or re-engagement as needed.)."*

2) **AI Prompt:** *Predictive Analytics for Cybersecurity Sales*
"Develop an AI-driven prompt to analyze cybersecurity purchase intent using segmentation variables like **[company size, industry, past security incidents]."***

3) **AI Prompt:** *AI Chatbot & Sales Assistant*
"Create a chatbot script for a cybersecurity website that qualifies leads by asking industry-specific questions and offering relevant resources before prompting a consultation booking."

Chapter 4: Sales Enablement – AI-Powered Sales Scripts & Call Guides

AI-Optimized Cold Calling & Lead Nurturing

1) **AI Prompt:** *High-Converting Cybersecurity Cold Call Script: "Develop a persuasive cold call script for [cybersecurity solution] targeting [specific company type or decision-maker]. Begin with a compelling opening that highlights a key statistic or threat, address common objections with data-driven rebuttals, and conclude with a clear call-to-action for a consultation or product demo. Emphasize loss aversion and urgency to prompt immediate action."*

2) **AI Prompt:** *Objection Handling for Cybersecurity Sales "Create responses for common cybersecurity objections, such as 'We already have security,' 'We don't have the budget,' and 'We're too small to be targeted.'"*

3) **AI Prompt:** *AI-Powered Lead Nurturing Sequence "Draft a three-step email sequence for nurturing cybersecurity leads. The emails should educate, provide value, and encourage the recipient to book a call."*

AI automation platforms streamline and optimize lead conversions, reducing manual effort while maximizing impact.

How It Works:

- AI Chatbots for Real-Time Engagement: AI-powered chatbots (e.g., Drift, Conversica) analyze visitor behavior and instantly suggest cybersecurity solutions tailored to their needs.
- Real-Time AI-Personalized Email Sequences: AI-driven platforms (e.g., Seventh Sense) analyze recipient

interaction history and send personalized follow-ups at the best possible time.

- Example: AI detects that a CISO attended a cybersecurity webinar but didn't book a call. AI then sends a follow-up email addressing webinar takeaways and inviting them to a one-on-one consultation.

Customizable AI Prompt: *"Generate an AI-powered follow-up email sequence targeting cybersecurity webinar attendees who engaged but didn't book a consultation. Include personalized recommendations and a CTA."*

Chapter 5: Building Thought Leadership & Influencer Collaboration

Establishing Brand Authority

1) **AI Prompt:** *Cybersecurity Thought Leadership Blog*
"Write a long-form blog on **[emerging cybersecurity trend]**, *targeting industry leaders with expert insights, statistics, and an engaging CTA."*

2) **AI Prompt:** *Podcast or Webinar Outline*
"Generate an outline for a webinar titled 'Future of Cybersecurity: Staying Ahead of Hackers in [2025],' including key discussion points and engagement strategies."

3) **AI Prompt:** *Influencer Partnership Outreach*
"Write an outreach email proposing a co-branded campaign with a cybersecurity influencer. Highlight collaboration benefits and include a CTA for a meeting."

Chapter 6: Gamification & Interactive AI Marketing

Engaging Cybersecurity Audiences

1) **AI Prompt:** *Gamified Cybersecurity Quiz*
"Create an interactive cybersecurity quiz titled 'How Hackable Are You?' featuring multiple-choice questions and a dynamic scoring system leading to a personalized security recommendation."

2) **AI Prompt:** *AI-Powered Social Media Giveaway*
"Develop a LinkedIn giveaway campaign where participants answer security-related trivia for a chance to win a **[security audit, software discount, book copy]."**

Chapter 7: Future-Proofing Your Cybersecurity Marketing Strategy

Anticipating Emerging Threats & Staying Ahead of Competitors

1) **AI Prompt:** *Expert-Level Thought Leadership Article* *"Write an expert-level thought leadership article on **[emerging cybersecurity threat: quantum computing, AI-powered phishing, supply chain attacks].** Include why it matters, potential risks, and how businesses can prepare."*

2) **AI Prompt:** *Text-to-Image AI Prompt for Visuals* *"Create a futuristic cybersecurity-themed image featuring **[a digital fortress, AI-powered security shields, quantum hacking defense],** representing the next evolution of cybersecurity in **[2025, 2030, beyond].** Use a mix of neon cyber aesthetics and ultra-realistic detail."*

Cybersecurity marketers can **leverage AI to detect real-time security threats** and immediately launch marketing campaigns that **address the latest risks.**

How It Works:

- AI-Powered Threat Feeds: AI tools like Recorded Future and Darktrace scan global security threats, predicting which vulnerabilities are most likely to be exploited.
- Real-Time Content Activation: AI marketing platforms like Pathmatics or Adzooma auto-launch targeted cybersecurity campaigns the moment a new attack method is identified.
- Example: AI detects a new zero-day exploit targeting Microsoft cloud infrastructure. AI then triggers an automated LinkedIn post, email campaign, and webinar alert addressing the new risk.

Customizable AI Prompt: *"Develop an AI-powered cybersecurity alert system that detects emerging threats and auto-generates marketing campaigns in response. Include [LinkedIn] posts, email sequences, and blog outlines."*

AI-generated content adapts in real-time to cybersecurity trends and breaking threats, allowing instantaneous campaign execution.

How It Works:

- Generative AI for Newsjacking: AI tools like Perplexity AI or ChatGPT scan industry news to instantly generate cybersecurity blog posts, LinkedIn updates, and video scripts.
- Example: AI detects a major supply chain attack and automatically creates a LinkedIn post with a call to action for Zero Trust solutions.
- Real-Time AI Video Generation: Tools like Synthesia allow companies to generate cybersecurity awareness videos with AI avatars within minutes.

Customizable AI Prompt: *"Generate a LinkedIn post and blog article on [latest cybersecurity incident] explaining how [cybersecurity solution] mitigates similar risks. Include a CTA for a free consultation.*

Chapter 8: The Psychology of Selling Cybersecurity.

Leveraging Fear, Trust & Urgency for Ethical Persuasion

1) AI Prompt: Persuasive Sales Pitch Script

"Create a persuasive sales pitch script that highlights the psychological impact of cyberattacks. Use the fear of financial loss, the power of trust, and the urgency of real-world case studies to compel decision-makers to act now."

2) AI Prompt: *Text-to-Image AI Prompt for Visuals*

*"Design a dramatic cybersecurity awareness poster that shows **[a breached network with a red 'ALERT' sign]**, evoking urgency and the consequences of ignoring cybersecurity."*

3) Enhanced Text-to-Image Prompt: *"Design a text-to-image prompt for generating visuals that align with your cybersecurity campaign. For example, create an image that embodies **[concept: futuristic cybersecurity, a digital fortress, or real-time threat detection]** using a sleek, professional aesthetic with **[specified color scheme or elements, e.g., a red 'ALERT' sign or live webinar screen]**. Adapt the prompt based on the campaign's focus."*

Chapter 9: Webinar Marketing – Driving Engagement & Conversions

Hosting High-Impact Webinars on Cybersecurity Topics

*1) **AI Prompt:** High-Converting Webinar Registration Page*
*"Generate a high-converting webinar registration page script for **[topic: ransomware defense, AI-driven security, Zero Trust implementation]**. Include an attention-grabbing headline, bullet-point benefits, and a clear call-to-action."*

*2) **AI Prompt:** Text-to-Image AI Prompt for Visuals*
*"Create a professional webinar promotional image featuring a **[cybersecurity expert speaking, a laptop screen displaying 'LIVE Webinar', and futuristic data security elements]**. Use a sleek tech-based style with a blue and dark cyber aesthetic."*

AI-generated video content and deepfake detection awareness campaigns are the next big frontier in cybersecurity marketing.

How It Works:

- AI Video Creation for Awareness: AI tools like Synthesia and RunwayML can generate cybersecurity training videos featuring AI avatars of real executives.
- Deepfake Detection Campaigns: AI-powered platforms (like Deepware or Truepic) can analyze fake cybersecurity threats and generate real-time awareness campaigns for customers.
- Example: A finance company detects a deepfake phishing attempt mimicking its CEO → AI auto-generates an awareness video alerting employees within minutes.

Customizable AI Prompt: *"Generate an AI-powered cybersecurity awareness video script highlighting deepfake phishing attacks, explaining how businesses can detect and prevent them."*

Chapter 10: Marketing Calendar & Event-Driven Campaigns

Aligning Marketing Efforts with Cybersecurity Awareness Events

1) **AI Prompt:** *Social Media Campaign for Cybersecurity Awareness Month*
"Write a social media campaign outline for **[Cybersecurity Awareness Month in October]**, *including daily themed posts, engaging challenges, and a call-to-action for businesses to take cybersecurity seriously."*

2) **AI Prompt:** *Text-to-Image AI Prompt for Visuals*
"Design a Cybersecurity Awareness Month infographic with engaging icons representing **[ransomware threats, strong passwords, phishing scams]**. *The aesthetic should be modern, professional, and educational."*

3) Event-Driven Campaign Prompt: "Generate an event-driven marketing campaign outline for [event name or holiday, e.g., National Password Day, Cybersecurity Awareness Month]. The campaign should include multi-channel content ideas—such as social media posts, email templates, and landing pages—that incorporate urgency and a clear call-to-action tailored to [target audience]."

Final Words: AI-Powered Cybersecurity Marketing – The Future is Here

AI prompt engineering is a transformative tool in cybersecurity marketing. By leveraging structured AI-generated content, businesses can enhance engagement, scale marketing efforts, and increase conversions. Implementing these prompts will empower cybersecurity professionals to drive impactful marketing campaigns and build stronger client relationships.

Quick-Start Guide for Cybersecurity Marketing Success

This Quick-Start Guide summarizes the key takeaways, AI prompts, and actionable strategies from each chapter to help you implement the ideas effectively.

Chapter 1: Leveraging Cyberattack Case Studies for Marketing and Sales

Key Takeaway:

Use real-world cyberattacks to educate prospects and demonstrate your solutions' relevance.

AI Prompt: *"Write a LinkedIn post targeting SMB owners about lessons learned from [cyberattack: ransomware, phishing]. Include actionable solutions and a CTA for a free vulnerability assessment."*

Action Item: Monitor cyberattack news (e.g., Google Alerts) and craft timely campaigns addressing specific vulnerabilities.

Chapter 2: Newsjacking and Emotional Storytelling

Key Takeaway: Turn breaking news into marketing opportunities and use emotional appeals to engage decision-makers.

AI Prompt: *"Create a marketing email targeting [industry: healthcare] with lessons learned from [recent breach]. Include emotional storytelling to highlight risks."*

Action Item: *Use tools like Feedly to track trending news stories and integrate them into your campaigns.*

Chapter 3: AI-Driven Marketing Strategies

Key Takeaway: Leverage AI to personalize campaigns, predict trends, and generate leads with unmatched precision.

AI Prompt: *"Generate a personalized email for [audience: SMB owners] offering a free consultation on [cybersecurity topic: ransomware prevention]."*

Action Item: Experiment with predictive analytics and chatbots to optimize lead qualification.

Chapter 4: Sales Enablement

Key Takeaway: Empower your sales team with tailored scripts and training to build trust with prospects.

AI Prompt: "Write a cold-call script targeting [industry: finance] focused on [solution: endpoint protection]." Action Item: Develop sales scripts that emphasize ROI and address specific pain points.

Chapter 5: Building Thought Leadership in Cybersecurity

Key Takeaway: Establish authority through high-value content, partnerships, and original research.

AI Prompt: *"Draft a blog post on [cybersecurity topic: Zero Trust] targeting SMBs. Include actionable steps and a CTA for a free assessment."*

Action Item: Publish educational content regularly, such as blog posts, webinars, or whitepapers.

Chapter 6: Future-Proofing Your Cybersecurity Marketing Strategy

Key Takeaway: Stay ahead of emerging threats by continuously monitoring industry trends and evolving your strategy.

AI Prompt: *"Analyze the top cybersecurity threats for [quarter/year] and recommend marketing strategies to address them."*
Action Item: Use A/B testing to refine campaigns and validate predictions.

Chapter 7: Psychology of Selling Cybersecurity

Key Takeaway: Use fear, trust, and urgency ethically to drive decision-making.

AI Prompt: *"Create a landing page targeting SMBs emphasizing trust and urgency to promote [solution: ransomware protection]."*

Action Item: Incorporate testimonials, certifications, and statistics to build credibility.

Chapter 8: Webinar Marketing

Key Takeaway: Use webinars to educate prospects, engage audiences, and generate leads.

AI Prompt: *"Draft a webinar agenda on [topic: ransomware protection] including live polls and interactive Q&A."*
Action Item: Host webinars on timely topics and promote them across multiple channels.

Chapter 9: Marketing Calendar and Event-Driven Campaigns

Key Takeaway: Align campaigns with national holidays and cybersecurity events for maximum relevance.

AI Prompt: *"Generate a campaign for [holiday: Data Privacy Day] targeting SMBs with a CTA for a free consultation."*
Action Item: Plan campaigns around key dates using a marketing calendar.

Chapter 10: Conclusion and Next Steps

Set clear KPIs (e.g., lead generation, ROI).
Launch campaigns tied to trends and metrics.
Train teams with ongoing workshops and AI tools.

Exclusive Resources for Marketing Success

https://911cybersecurity.com/book/

Acknowledgments Section

Finally, I want to extend my deepest gratitude to my wife. Her unwavering support, encouragement, and belief in everything I do have been the foundation of my life. Her strength and wisdom inspire me daily, and this book would not have been possible without her unconditional help and understanding.

Thank you for being my partner in all aspects of life.

Recommended Resources More Resources for Cybersecurity Success

Complimentary Book Resources Hub
To help you maximize the value of this book, I've created a resource hub on 911Cybersecurity.com/book/

The hub includes:

- Bonus templates and checklists
- Ready-to-use AI prompts
- Access to updated case studies
- Webinars and video tutorials for deeper insights

Technology Marketing Toolkit by Robin Robins,
Robin Robins is a highly respected authority in marketing for MSPs and IT service companies. Her company, Technology Marketing Toolkit, provides proven systems, campaigns, and resources that help small MSPs achieve outstanding growth in competitive markets. From detailed strategies to ready-to-use templates, Robin's advice has helped thousands of IT companies grow their revenue and client base.

Visit their website:
https://www.technologymarketingtoolkit.com/

I highly recommend exploring their content and resources to complement the strategies outlined in this book. While we are not affiliated, this recommendation reflects my respect for their expertise and the actionable value they bring to the industry.

"The Psychology of Marketing - Unlock the Secrets of the Human Mind to Revolutionize Your Marketing Strategy". by Mrs. Melody Thompson (Author), Mr. Nikolay Gul (Illustrator)

Credit and Acknowledgments

This book features real-world examples, insights, and inspirations drawn from respected publications, companies, and thought leaders. I want to extend my gratitude to the following:

Industry Case Studies and Examples

If any real companies are referenced in this book, their contributions to the field of cybersecurity are fully acknowledged as benchmarks and inspirations for marketing innovation. All mentions are used respectfully and non-affiliatively.

If you believe I have overlooked any credits, please do not hesitate to contact me via my LinkedIn profile or at 911Cybersecurity.com to address this.

Notes on Borrowed Ideas

This book is the result of years of research, practical experience, and creativity. While original ideas form its core, certain inspirations are derived from widely respected publications, online articles, and thought leaders in the field of cybersecurity and marketing. Whenever applicable, links have been provided to direct you to the original sources. If there are uncredited references, they are unintentional, and I will gladly acknowledge them upon notification.

Use of AI.

Just like many writers have used ghostwriters for years (it's more common than you think – some studies suggest up to 80% of non-fiction books are ghostwritten in some capacity), AI is now a tool that can be used effectively.

Final Notes and Call to Action

Your journey as a cybersecurity marketer doesn't end here. Use the strategies in this book to improve your marketing campaigns, drive sales, and build lasting trust with your clients. I encourage you to:

- Share your results and feedback with me on LinkedIn: https://www.linkedin.com/in/webdesignerny/
- Explore additional insights and tools at 911Cybersecurity.com/book/

Thank you for trusting me as your guide in this ever-evolving landscape of cybersecurity marketing. Here's to your success!

Let's make your mark and enhance the future of high-tech and cybersecurity marketing!

Legal Disclaimer

This book is intended as an educational resource for marketing and sales professionals in cybersecurity and high-tech industries. The strategies, frameworks, AI prompts, and methodologies presented are original unless explicitly cited. The author does not claim ownership of referenced business strategies, books, trademarks, or brand names.

All references to publicly known books, frameworks, and companies—including but not limited to The Art of War, The 21 Immutable Laws of Marketing, major cybersecurity vendors, or tech brands—are made for educational, illustrative, or comparative purposes only. The author and publisher acknowledge all original authors and copyright holders of cited materials.

The content in this book is provided **"as is"** without warranties or guarantees of success. Readers are responsible for **adapting and verifying** marketing strategies to fit their business model, industry regulations, and legal requirements. The author and publisher disclaim any liability for outcomes—intended or unintended—resulting from the use of information contained within this book.

Readers who purchase this book are welcome to apply the strategies, prompts, and ideas to their own business efforts. However, redistribution, resale, or reproduction of this content for commercial use without permission is prohibited.

This book is an **independent work** and is **not affiliated with, endorsed by, or officially associated with** any company, brand, or entity mentioned.

By reading this book, you acknowledge and agree that **implementation is at your own discretion and risk.**

"Future-Proof Cybersecurity Marketing: Dominate with AI-Driven Strategies" by Nikolay Gul

New York, USA
911CyberSecurity.com/book

Connect with the author, Nikolay Gul,
on LinkedIn for more insights and updates:
https://www.linkedin.com/in/webdesignerny/

LCCN (Library of Congress Control Number): **2025902819**
ISBN: 9798218612481

127

www.ingramcontent.com/pod-product-compliance
Lightning Source LLC
Chambersburg PA
CBHW071710210326
41597CB00017B/2415